Contents

Revising for the Standard Grade Physics course

Syllabus

The Standard Grade Physics course covers the following 7 topics:

1 Telecommunications
2 Using Electricity
3 Health Physics
4 Electronics
5 Transport
6 Energy Matters
7 Space Physics

Assessment

There are two types of assessment – **external** and **internal.**

The **external assessment** consists of two separate examination papers which are designed to assess the elements of **Knowledge and Understanding (KU)** and **Problem Solving (PS)**:

1 **General** paper which assesses Grades 3 and 4 and lasts for 1 hour and 30 minutes.

2 **Credit** paper which assesses Grades 1 and 2 and lasts for 1 hour and 45 minutes.

The **KU** and **PS** elements are graded separately in each paper. If you attempt papers at both Credit and General levels, you will be awarded the better of the two grades achieved in the two papers for each element. Grade 5 may be awarded to candidates who narrowly fail to meet the criteria for General Level.

Practical Abilities (PA) which consist of techniques and investigations, are **assessed internally** and you will be awarded a grade in **PA** based on your performance.

Your grade for attainment in each of the elements **KU, PS** and **PA** will be recorded on your final Certificate together with an overall grade. The overall grade is worked out by taking an average of your grades in KU, PS and PA with a weighting of 2:2:1 in favour of KU and PS.

Structure and aim of this book

The aim of this book is to help you achieve success in the final exam through understanding fundamental principles and processes in Physics. It provides you with a concise coverage of the syllabus content, a selection of the most frequently occurring problems posed from questions in past papers and advice on how to prepare for the SQA exam. Many of the Credit questions contain material covering the General Learning Outcomes so Credit candidates really should be comfortable with all of the General and Credit questions in this book when they approach the May exam.

The book is divided into the 7 topics of the course and each topic:

▶ covers the learning outcomes at General and Credit levels.

continued

BrightRED Results

Standard Grade
PHYSICS

Sarah Fletcher, Andrew McGuigan,
Andy Shield

First published in 2011 by:

Bright Red Publishing Ltd
6 Stafford Street
Edinburgh
EH3 7AU

A CIP record for this book is available from the British Library

ISBN 978-1-906736-07-1

With thanks to Ken Vail Graphic Design, Cambridge (layout) and Tony Wayte (copy-edit)

Cover design by Caleb Rutherford – eidetic

Illustrations by Ken Vail Graphic Design, Cambridge.

Acknowledgements

Every effort has been made to seek all copyright holders. If any have been overlooked then Bright Red Publishing will be delighted to make the necessary arrangements.

Bright Red Publishing would like to thank the Scottish Qualifications Authority for use of Past Exam Questions. Answers do not emanate from SQA.

Bright Red Publishing would like to thank the following for permission to reproduce the following photographs:

© Hornby (p 25)
© istockphoto (pp 21, 25, 29, 65)
© Shutterstock (pp 29, 61, 113)
© NASA (p 113)

Printed and bound in Scotland by Bell & Bain Limited, Glasgow

Mixed Sources
Product group from well-managed forests and other controlled sources
www.fsc.org Cert no. TT-COC-002769
© 1996 Forest Stewardship Council
FSC

Structure and aim of this book – continued

▶ contains General and Credit level questions taken or adapted from past papers with detailed explanations and guidance in providing answers.

▶ contains **Look out for!** sections and hints where you are reminded of common problems and difficult aspects of the course are explained.

Full details of the course arrangements are available from the SQA website at www.sqa.org.uk.

Formulae

Become familiar with using Page 3 in the SQA Data Booklet provided in exams:

▶ Do not use the equations needed for Int 2, i.e. $s = vt$, $p = mv$

▶ Learn the meanings of the symbols

▶ Practise rearranging the equations

▶ Learn the units associated with each of the quantities

▶ Refer to the glossary for explanations and for symbols and units

Look out for

This booklet is also published on the SQA website. Make sure you are familiar with the symbols and units required for each equation.

Look out for

$p = mv$ not needed for SG

Look out for

V gain and P gain require no units

The data sheet

This is published at the beginning of the Credit Paper. Any necessary data which you need for General Level will be given in the question.

Look out for

Become familiar with the data and the most common data needed i.e. the speeds of light and sound in air, the gravitational field strength of Earth, the specific heat capacity of water etc.

Look out for

Do not confuse the speed of sound 340 m/s with the speed of light 3×10^8 m/s! This is also the speed of all the waves in the electromagnetic spectrum (radio, microwaves infra-red etc)

Look out for

Do not confuse *fusion* (liquid to solid) with *vaporisation* (liquid to gas)

Look out for

Be careful not to confuse: m – milli ($\frac{1}{1000}$ or 10^{-3}) with M – mega (10^6)

Communication using waves

The speed of sound

What you should know at **General** **and** **Credit** **level...**

The speed of sound in air is less than the speed of light in air. Examples which illustrate this fact include firing a starter's pistol, where the sound is heard after seeing the smoke and in a storm, thunder is heard after seeing the lightning. Speed, distance and time are related to each other in calculations involving problems on sound transmission and water waves.

Look out for

Remember – the speed of light is approximately 1 million times faster than the speed of sound in air.

Speed is defined as the distance travelled in one second by a moving object.

$$speed = \frac{distance}{time} \quad or \quad v = \frac{d}{t},$$ where v = speed in m/s
d = distance in m
t = time in s.

Look out for

The symbol for seconds is **s**, not **secs**.

Speed calculations also occur in the topics on Health Physics, Transport, Energy Matters and Space Physics, so the formula for speed is very important. Make sure you practise manipulating it to find distance and time:

$$d = vt \qquad t = \frac{d}{v}$$

Speed of sound in air (as quoted in the data sheet) = 340 m/s.

The speed of sound depends on the material it is passing through, for example:

▶ speed of sound in water = 1500 m/s
▶ speed of sound in steel = 5200 m/s

A table of the speed of sound in materials is included in the data sheet provided in the Standard Grade Credit Exam. You will be given any necessary data in General questions.

General question 1

An onlooker watching an old chimney being demolished sees the puffs of smoke and dust from the explosion. She hears the bang of the explosion 6 seconds later.
(a) Explain why there is a delay between the onlooker seeing the smoke and hearing the explosion.
(b) Calculate the distance between the onlooker and the explosion. (The speed of sound in air is 340 metres per second.)

General question 1 – Answer

(a) _The speed of light (from the puffs of smoke and dust) is much greater (300 000 000 m/s) than the speed of sound (from the explosion)._

(b) _$d = v \times t$_
     _= 340 × 6 = 2040 m_

The speed of sound is given in General questions. In Credit questions you will have to get it from the data sheet.

Credit question 1

A student sets up the apparatus **exactly** as shown to measure the speed of sound in air.

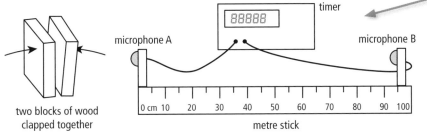

Clapping the two blocks of wood produces a sound. Timing starts when the sound reaches microphone A and stops when the same sound reaches microphone B.

The time taken for sound to travel between the two microphones is recorded as 2·85 ms. Using information from the diagram, calculate the speed of sound.

It is important to note that the microphones are 0·95 m apart (always check the diagram carefully if asked to use information from it). Also be careful when converting from ms (milliseconds) to s (seconds).

Round up this answer to 333 m/s. You will lose ½ mark for using too many significant figures.

Credit question 1 – Answer

$$speed = \frac{distance}{time} = \frac{0·95}{0·00285} = 333·333 \, m/s = 333 \, m/s$$

Look out for

Do not assume that the value will be the same as quoted in the data sheet. This is an experiment which can give varying results.

Waves

What you should know at General level...

All waves can be defined by a set of related properties: **frequency**, **wavelength**, **speed**, **energy**, and **amplitude**.

Speed, wavelength and frequency are related to each other in calculations using problems on water and sound.

Waves carry energy from one place to another and can be used to transmit signals. Sound, light and water ripples in the sea are examples of waves.

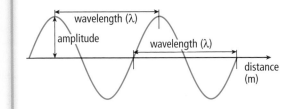

Quantity	Definition	Unit
wavelength	the shortest distance before the wave pattern repeats	metres, m
frequency	number of waves that pass a point in one second	**hertz**, Hz
speed	the distance travelled in unit time	metres per second, m/s
amplitude	distance from the rest position to top of a crest or the bottom of a trough	metres, m

Look out for

Learn these definitions and units. Refer to the glossary.

continued

What you should know at General level – continued

The frequency f is the number of waves which pass a point in one second. Each wave has a wavelength, given by the Greek letter lambda, λ and is measured in metres. The speed of the wave v is the distance travelled by the wave in one second and is calculated by multiplying the frequency by the wavelength:

$$v = f\lambda$$

Look out for

This is another important equation. Make sure you practise manipulating it i.e. $f = \frac{v}{\lambda}$, $\lambda = \frac{v}{f}$.

What you should know at **Credit** level...

The total distance travelled by a wave in one second is found by multiplying the frequency by the wavelength:

distance d travelled in one second $= f \times \lambda$

Knowing the distance travelled in one second lets us work out the velocity of a wave in terms of its frequency and wavelength:

the velocity v of a wave is given by $v = \dfrac{d}{t}$

this is the same as $v = \dfrac{f\lambda}{t}$. If $t = 1$, this becomes $v = f\lambda$

This means we have equivalent equations for velocity:

$v = \dfrac{d}{t}$ and $v = f\lambda$

Communication using cables
Signals in wires

What you should know at **General** level...

Coded messages or signals (such as Morse code or similar) can be sent out by a transmitter and are picked up by a receiver. Morse code uses a series of long and short pulses of energy, which could be sound or light.

The telephone is an example of long range communication using wires between transmitter and receiver.

Electrical signals travel along a wire at almost the speed of light 300 000 000 m/s. This can also be written as 3×10^8 m/s.

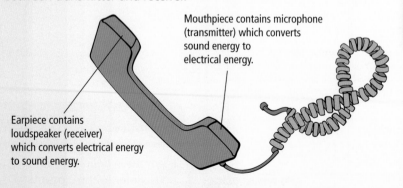

Mouthpiece contains microphone (transmitter) which converts sound energy to electrical energy.

Earpiece contains loudspeaker (receiver) which converts electrical energy to sound energy.

Look out for

This is a fact which you must know. Remember the units – if you don't use the units, you will score 0 marks.

What you should know at Credit **level...**

When a microphone is connected to an **oscilloscope** (an instrument which converts electrical signals into wave patterns on a screen), the pattern of sound waves can be examined.

When the loudness and frequency of a sound are adjusted, the patterns on the oscilloscope change, as shown in the diagrams:

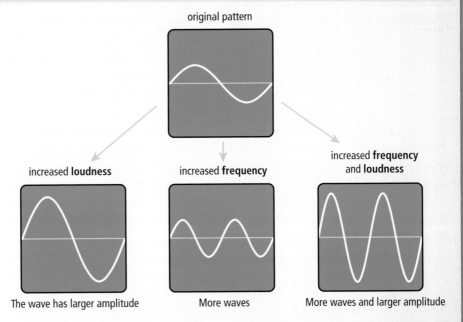

original pattern

increased **loudness**

increased **frequency**

increased **frequency** and **loudness**

The wave has larger amplitude

More waves

More waves and larger amplitude

General question 1

When a student whistles a note into a microphone connected to an oscilloscope, the following pattern is displayed:

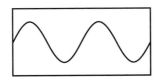

Without changing the oscilloscope controls, another student whistles a quieter note of higher frequency into the microphone.
(a) Which of the following shows the pattern which would be displayed on the screen?
(b) What sounds would have produced the other wave patterns?

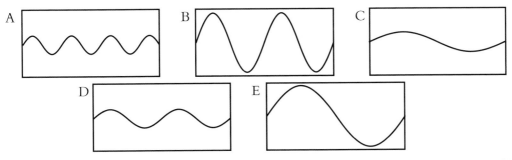

A

B

C

D

E

General question 1 – Answer

(a) *A – smaller pattern with more waves*
(b) *B – louder sound with same frequency   C – quieter sound with lower frequency   D – quieter sound with same frequency*
E – louder sound with lower frequency

Optical fibre communications

An optical fibre is a thin clear thread of glass along which light can travel. When electrical signals are converted into light using a laser or LED, the fibre carries the signals at about 2×10^8 m/s. Telephone, cable TV systems and computer networks can use optical fibres instead of electrical cables.

Reflection

Optical fibres use the basic property of reflection. When light meets a reflecting surface such as a mirror, the direction of the reflected light (**reflected ray**) is determined by the direction of the incoming light (**incident ray**). The law of reflection states:

the **angle of incidence** = the angle of reflection

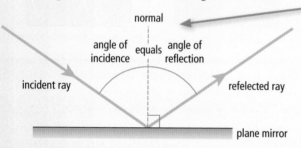

The **principle of reversibility** of ray paths states that if light is directed backwards along the path of the reflected ray, it will be reflected along the same path as the original incident ray.

In light ray diagrams, angles are always measured between the ray and the **normal**. *The normal is the perpendicular line from the mirror at the point where the incident ray meets the boundary.*

**Look out for**

Reflection is also the basic principle upon which curved reflectors in satellite receivers operate. See page 18.

Credit question 1

A farm road joins a main road at a bend. The farmer has placed a mirror as shown so that he can see when cars are approaching.

(a) Draw rays on the diagram to show how the farmer can see the car using the mirror. You must label the angles of incidence and reflection.

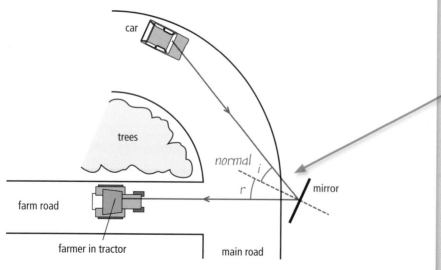

Remember all angles must be measured to the normal and use a ruler when drawing ray diagrams!

(b) State why the driver can also see the tractor using the mirror.

Light ray paths are reversible.

Total internal reflection

What you should know at **General** **level...**

Optical fibres use a property called **total internal reflection** (**TIR**). When light is shone inside a piece of glass, some part of the light is reflected at the inside surface. The amount of light that is reflected depends on the angle of incidence.

Every material has a **critical angle**:

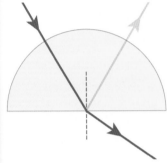 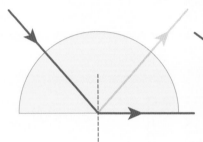 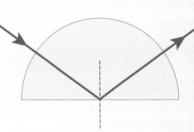

▶ if the angle of the incident ray inside the glass is **less** than the critical angle, the light is **refracted** (passes out of the glass at a slightly different angle to the incident angle)

▶ if the angle of the incident ray inside the glass is **equal** to the critical angle, the light is transmitted along the edge of the glass

▶ if the angle of the incident ray inside the glass is **greater** than the critical angle, all the light is **reflected** inside the glass. This is total internal reflection.

In glass, the critical angle is about 42°.

Light passing down an optical fibre

In an optical fibre, light travelling along the fibre always undergoes total internal reflection at the boundary between the fibre core and cladding.

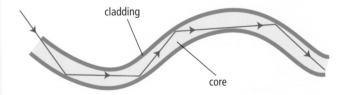

cladding

core

To send an electrical signal it must be converted by a transmitter to light. The light is transmitted along the optic fibre and then is changed back to an electrical signal at the receiver.

Look out for

If you have to draw a diagram to show the transmission of light in an optical fibre, avoid drawing an excessive number of reflections. Make the reflections realistic i.e. angles 'i' and 'r' should be equal. Use a ruler – if one is not available make the lines as straight as possible!

Look out for

Remember – in calculations light travels in glass at 2×10^8 m/s. This is in the data sheet.

Comparing electrical cables and optical fibres

What you should know at **Credit** level...

Optical fibres and electrical cables can both be used in telecommunications systems. Optical fibres have a number of advantages over electrical cables:

▸ optical fibres are smaller, lighter and cheaper than electrical cables

▸ more signals can be sent in a much smaller cable

▸ signal quality and capacity are higher in optical fibres

▸ no electrical interference

▸ cannot be easily tapped

▸ less energy is lost so less amplification is needed. This is measured as signal reduction per kilometre, and is about 25 times better for optical fibre systems.

Credit question 1

A computer is connected to the internet by means of a copper wire and a glass optical fibre as shown.

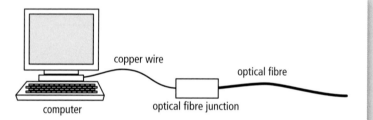

(a) In the table, enter:
   (i) the speed of the signal in each material (ii) the type of signal in each material.
(b) Complete the diagram to show how the signal travels along the optical fibre.
Copper wire or glass optical fibre can be used in telecommunication systems.
(c) Explain which material, copper or glass, would need fewer repeater amplifiers over a long distance.
(d) A broadband communications system carries over 100 television channels and 200 phone channels. Explain which material, copper or glass, should be used in this system.

Credit question 1 – Answer

(a)

	Copper wire	Glass optical fibre
Speed of signal	3×10^8 m/s	2×10^8 m/s
Type of signal	electrical	light

(b)

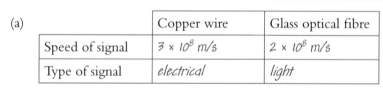

(c) _Glass. One of the advantages of using fibre optics is that there is less energy lost._
(d) _Glass. Fibre optics can carry much more information and channels._

Signal transmission

What you should know at Credit **level...**

In telecommunication systems, electrical signals are converted to light signals using a **laser** or **LED** to be transmitted through optical fibres. The light signals are then converted back into electrical signals.

Credit question 1

A laptop computer uses a radio signal to transfer information to a base station. The base station is connected by optical fibres to a telephone exchange.

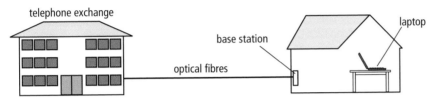

telephone exchange

base station

laptop

optical fibres

The telephone exchange is 40 km away from the base station.

Calculate the time taken for the signal to travel along the **glass** optical fibre from the base station to the local telephone exchange. (3)

Credit question 1 – Answer

$v = 2 \times 10^8 \, m/s$

$t = \dfrac{d}{v}$

$= \dfrac{40\ 000}{2 \times 10^8}$

$= 2 \times 10^{-4} \, s \ (0 \cdot 0002 \, s)$

One mark for selecting the speed of light in glass.

Look out for

When an answer is worth 3 marks there is usually an additional piece of information required to the standard 2 mark calculation.

Radio and television

Parts of a radio receiver

What you should know at General **level...**

The main parts of a radio receiver are the aerial, tuner, decoder, amplifier, loudspeaker and electricity supply. These are shown on the diagram.

Aerial – collects all the waves and converts them to small electric currents.

Tuner – selects one radio signal or frequency.

Decoder – separates the higher frequency **carrier wave** from the lower frequency **audio signal**.

Amplifier – increases the electrical energy to make it large enough to operate the loudspeaker.

Loudspeaker – changes the electrical energy into sound energy.

Electricity supply – provides energy for the amplifier.

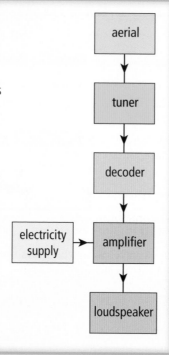

Look out for

These are commonly asked so make sure you learn and understand the parts and their functions.

Look out for

Block diagrams – the electricity supply is sometimes not included in questions.

Modulation

What you should know at Credit **level...**

In order to carry information (speech or music for example), a radio wave must be **modulated** (or changed) in some way.

The following two diagrams represent a sound wave and a radio wave.

Audio wave produced by a sound signal

Radio or carrier wave

Modulated wave

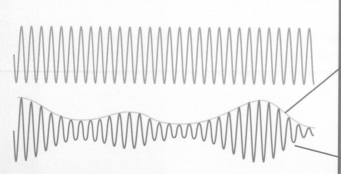

Look out for

Notice the relatively low frequency of the audio wave compared to that of the radio wave.

Look out for

The shape of the envelope of the wave corresponds to the shape of the audio wave.

Look out for

The shape of the inner wave corresponds to the frequency of the carrier wave.

Television

What you should know at **General** level...

The main parts of a television receiver are the aerial, tuner, decoders, amplifiers, tube, loudspeaker and electricity supply as shown.

The tuner operates in a similar way to the radio receiver with additional parts to separate the audio signal from the video signal.

```
                                              electricity
                                                supply
                                                  │   │
                                                  ▼   ▼
                        ┌──────────┐     ┌──────────┐     ┌──────────┐
                   ┌───▶│  vision  │────▶│  vision  │────▶│ picture  │
                   │    │ decoder  │     │amplifier │     │   tube   │
┌────────┐   ┌────────┐ └──────────┘     └──────────┘     └──────────┘
│ aerial │──▶│ tuner  │
└────────┘   └────────┘ ┌──────────┐     ┌──────────┐     ┌──────────┐
                   └───▶│  audio   │────▶│  audio   │────▶│loudspeaker│
                        │ decoder  │     │amplifier │     └──────────┘
                        └──────────┘     └──────────┘
```

Aerial – collects all the waves and converts them to small electric currents.

Tuner – selects one radio signal or frequency.

Vision decoder – separates the video signal from the higher frequency carrier wave.

Audio decoder – separates the audio signal from the higher frequency carrier wave.

Amplifiers – increase the audio and video signals to make them large enough to drive the loudspeaker and the screen.

Loudspeaker – changes the amplified electrical audio signal energy into sound energy.

Screen – changes the amplified video electrical energy into light energy.

Electricity supply – provides energy for the decoders and amplifiers.

General question 1

(a) Which part of a television receiver picks up the incoming signal?

(b) What is the purpose of the electricity supply in a television receiver?

General question 1 – Answer

(a) *aerial*

(b) *To provide the vision and audio amplifiers with electrical power or energy.*

How a picture is produced on a TV screen

What you should know at General **level...**

The electron beam scans the screen in lines. This is a bit like reading a newspaper. Starting at the top of the screen, the beam moves from left to right, then back and down slightly. When the beam of electrons hit the screen their energy is converted into light by the coating on the screen. One picture is formed by 625 lines.

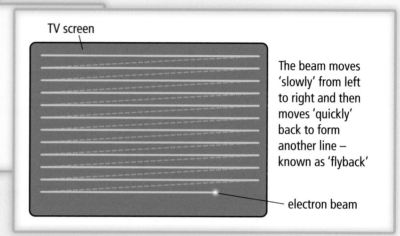

TV screen

The beam moves 'slowly' from left to right and then moves 'quickly' back to form another line – known as 'flyback'

electron beam

What you should know at Credit **level...**

The moving pictures you see on a TV screen are actually many pictures produced every second. The eye takes about 0·1 s to become aware of an object in front of it. The vision persists for about 0·1 s after the object is removed. When the brain receives pictures quickly one after another, it holds each picture for a short time. If the next image appears quickly enough, the brain is not aware of the space and they merge into each other.

The eye and brain cannot separate all the different pictures because of **image retention** so the pictures merge into each other.

Brighter areas are produced by varying the number of the electrons in the electron beam. The more electrons, the brighter the spot.

Credit question 1

A door entry system in an office block allows video and audio information to be sent between two people. The system uses a black and white television screen. Describe how a moving picture is seen on the screen. Your description must include the terms:

line build up   image retention   brightness variation   (3)

Credit question 1 – Answer

A TV picture is composed of a series of lines on the screen. Each line is built up by electrons scanning across the screen. Brightness variation of each line is caused by changing the number of electrons hitting the screen. Each picture is slightly different from the previous picture and image retention by the brain causes the pictures to merge together giving the impression of movement.

Make sure your descriptions are full and clear.

The colour television

What you should know at **General** level...

All of the colours on a television screen are produced by mixing the three primary colours in various combinations.

The three primary colours are: red, blue and green. Other colours are produced by combining light as shown.

What you should know at **Credit** level...

In a colour TV there are 3 electron guns (instead of 1 gun in a black and white TV.) These fire colourless electrons at sets of 3 different coloured spots on the screen which glow when hit. Each electron beam hits one of the dots to produce the colour directed by a shadow mask which is a metal grid. There are no coloured electrons! If a red and blue dot are hit at the same time, the resulting colour is magenta.

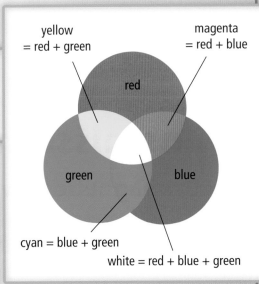

yellow = red + green

magenta = red + blue

red

green

blue

cyan = blue + green

white = red + blue + green

General question 1

The three colours of light that mix to produce all the colours seen on a TV screen are:

A   red, green and blue
B   red, yellow and blue
C   magenta, green and cyan
D   magenta, yellow and cyan
E   green, yellow and blue

General question 1 – Answer

Answer A

Look out for

Make sure you know what colour is produced by combining different colours.

Credit question 1

In a colour television tube, three electron guns each send a beam of electrons to the screen.

(a) Why are **three** electron guns needed in a **colour** television tube? (1)

(b) The diagram shows the screen and the shadow mask in a colour television screen. Use information from the diagram to explain why a shadow mask is needed. (2)

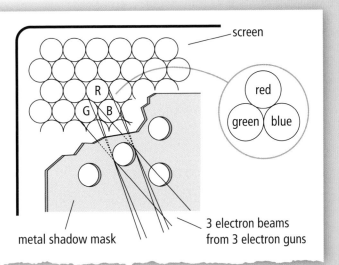

screen

R

G B

red

green blue

metal shadow mask

3 electron beams from 3 electron guns

Credit question 1 – Answer

(a) *Each gun aims electrons at a particular colour dot on the TV screen (i.e. blue, green and red).*

(b) *This is a metal grid which directs the electron guns to their particular coloured dots (phosphors).*

Transmission of radio waves

Electromagnetic waves

What you should know at **General** **level...**

Microwaves, television and radio signals are **electromagnetic waves** which carry energy through air or space at the speed of light (300 000 000 m/s) or 3×10^8 m/s without wires between the transmitter and receiver.

A transmitter can be identified by wavelength or frequency values.

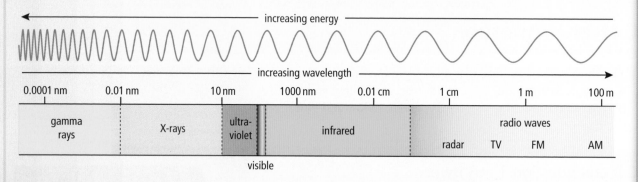

Look out for

Important! You will come across the EM spectrum again in the Health and Space Physics Units.

Look out for

We can tune to radio stations using their particular frequency or wavelength.

What you should know at **Credit** **level...**

If the frequency of a TV or radio station is known, it is always possible to calculate the wavelength, using the equation:

$\lambda = \frac{f}{v}$

Remember the speed of the waves is 300 000 000 m/s.

The range of electromagnetic waves depends on the wavelength. This affects how signals are reflected and refracted in the atmosphere. Some long wave **radio waves** are reflected by layers of charged particles very high in the atmosphere.

continued

Diffraction

Diffraction is the bending of waves around obstacles such as buildings and hills.

Long waves diffract more than short wavelengths so reception of long waves is better in hilly areas.

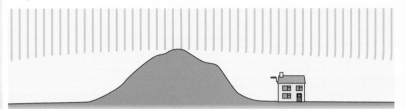

In a hilly region, the short waves used for TV transmission do not reach the house. TV reception will not be available.

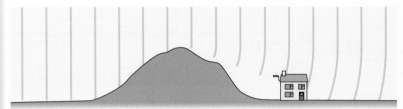

The longer waves (e.g. radio waves) reach the house.

Look out for

Important – this must not be confused with *refraction* which occurs when waves travel from one material to another. Carefully learn their definitions and their spelling!

General question 1

The frequency range and some uses of different radio wavebands are shown.

Waveband	Frequency range (megahertz)	Uses
HF	3 to 30	amateur radio, military communications
VHF	30 to 300	FM radio, air traffic control
UHF	300 to 3000	radar, local TV
SHF	3000 to 30 000	satellite TV, microwave communication

(a) Give a use, **from the table**, for a radio wave which has a frequency of 106 megahertz.   (1)

(b) TV is broadcast in the United Kingdom on the UHF band. What is the range of frequencies in this waveband?   (1)

General question 1 – Answer

(a) *FM radio or air traffic control.*

(b) *300 to 3000 megahertz*

Credit question 1

A hill lies between a radio and television transmitter and a house.

The house is within the range of both the radio and the television signals from the transmitter.

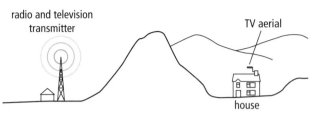

In the house, a radio has good reception but a TV has poor reception from this transmitter. Suggest an explanation for this. (2)

Credit question 1 – Answer

Radio waves have a longer wavelength than TV waves. (1)

Longer waves diffract more. (1)

Look out for

2 marks = 2 points in the answer.

Dish aerials and curved reflectors

What you should know at General **level...**

Curved reflectors on certain aerials or receivers make the received signal stronger. Curved reflectors are often used in telecommunication systems such as satellite TV, TV link, boosters, repeaters or satellite communication.

A **dish aerial** can be used to either **receive** or to **transmit** a radio (or microwave) signal.

This is how satellite television is received in our homes.

The receiver is placed at the focus of the curved reflector. This concentrates the signal which improves the reception.

Careful – it does not amplify the signal.

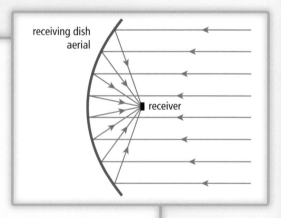

What you should know at Credit **level...**

Curved reflectors can be used to improve the transmitted signal.

The transmitter is placed at the focus of the curved reflector so that the waves are sent out in a parallel beam which is stronger than a beam which is emitted in all directions.

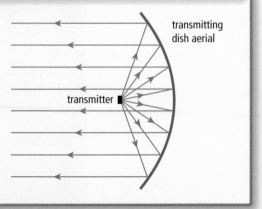

Look out for

If asked to complete a diagram as above: use a ruler to draw straight lines and make sure you have read whether it is a transmitter (signal going out) or a receiver (incoming signal). Remember arrows.

Satellites

What you should know at **General** and **Credit** **level...**

Satellites orbiting the Earth are used to transmit TV and radio signals. The time taken for a satellite to go round the Earth depends on the height of the orbit above the Earth. The **period** of a satellite is the time taken for it to complete a rotation around the Earth. The further away it is from Earth the longer the period. A **geostationary satellite** is at a height where a complete orbit of Earth takes 24 hours so it remains above the same point on the Earth's surface.

General question 1

State what is meant by a geostationary orbit.

General question 1 – Answer

The period is 24 hours
or *It is always above the same point on the Earth*
or *It has the same period as the Earth.*

Intercontinental communication

Radio waves cannot travel directly to distant receivers on Earth because of the curvature of the surface. The signals are transmitted to a satellite which re-transmits the signal to a receiver on the ground.

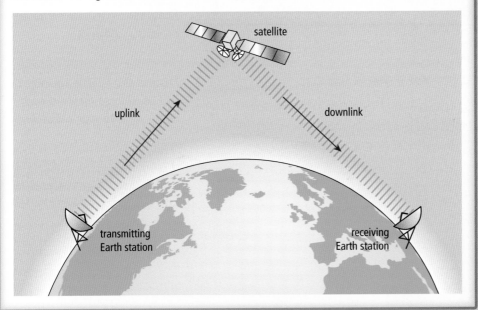

Credit question 1

A television company is making a programme in China. Britain receives television pictures live from China. The television signals are transmitted using microwaves. The microwave signals travel from China via a satellite, which is in geostationary orbit.

(a) State what is meant by geostationary orbit. (1)

(b) The diagram shows the position of the transmitter and receiver. Complete the diagram to show the path of the microwave signals from China to Britain. (2)

The frequency of the microwave signals is 8 GHz.

(c) What is the speed of the microwaves? (1)

(d) Calculate the wavelength of these microwaves. (2)

Remember prefixes are in the data sheet. G = giga = 10^9

Credit question 1 – Answer

(a) *The period is 24 hours or period same at the Earth.*

Do not write 'the same speed as the Earth'.

(b)

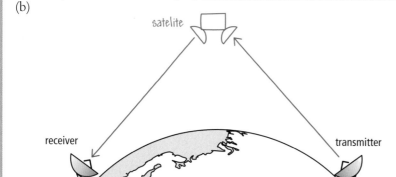

satelite

receiver   transmitter

Britain   China

(c) 3×10^8 m/s

Remember to include the units. no units = 0 marks

(d) $\lambda = \dfrac{v}{f}$

$= \dfrac{3 \times 10^8}{8 \times 10^9}$

$= 0.0375$ m

Practise rearranging equations and using large numbers such as 3×10^8 into your calculator.

From the wall socket

Household appliances

What you should know at **General** and **Credit** level...

All electrical circuits need a supply of energy. Supplies of electrical energy include **batteries** and the **mains supply**. A battery is a **d.c.** supply; d.c. stands for **direct current**, where the charges always move in the same direction round the circuit. The mains supply is **a.c.**; a.c. stands for **alternating current**, where the direction of movement of the charges alternates between the terminals of the supply. The declared value of voltage for the mains supply in Britain is 230 V and the frequency is 50 Hz. The declared value of the mains voltage is less than its peak voltage (325 V).

Electrical power and energy

Household appliances transform **electrical energy** into other useful forms of energy. The rate at which an appliance transforms energy is called its **power**. Electrical appliances have information about their power rating and voltage listed on a **rating plate**. The power rating is used to determine the correct **fuse** to fit in the plug in order to protect the flex. Appliances with a higher power rating carry more current so the wires in the flex are thicker. Fitting too low a value of fuse will result in the fuse melting every time the appliance is switched on. If the fuse value is too high and a fault develops the fuse may not break and the flex may overheat. Similarly it is important to fit the correct thickness of flex. Fitting a thin flex to an appliance with a high power rating will cause the flex to overheat and possibly catch fire.

This appliance works for a range of voltages but remember the declared value of the mains in Britain is 230 V

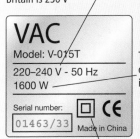

The power rating of the appliance is shown here

This symbol is the **Double Insulation** symbol. This means that the appliance uses 2-core cable with live and neutral wires, and does not require an earth wire.

General question 1

State the **useful** energy change for (a) a lamp and (b) a kettle.

General question 1 – Answer

(a) Lamp: Electrical energy → Light energy
(b) Kettle: Electrical energy → Heat energy

The question asks for the **useful** energy change, so think about the **purpose** of the appliance. A lamp will also give off heat, but that wouldn't be a useful energy change. Similarly, you might hear sound from your kettle but that also wouldn't be a useful energy change.

Look out for

In exam questions, markers will accept **electrical energy** or **electric energy** as the input energy but you will not get any marks for calling it **electricity** or **electricity energy**. You need to be precise with the language you use in a Physics exam.

Electrical wiring and insulation

What you should know at **General** and **Credit** level...

Many appliances have flexes which contain three wires; live, neutral and earth.

The insulation on these wires is colour coded:

▶ live   brown

▶ neutral   blue

▶ earth   green and yellow

It is important that the conductors are connected to the correct connections in the plug, otherwise the appliance may not work or may be dangerous to use.

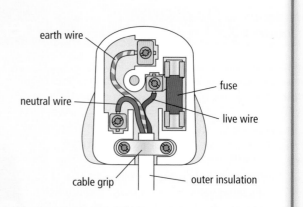

The switch and the **fuse** must be connected in the live wire, so that the appliance can be isolated from the mains supply. If the switch was in the neutral wire, then the appliance would still be connected to the live wire even if the switch was open. Similarly, if the fuse was connected in the neutral wire and a fault developed so that the fuse melted, the appliance would still be connected to the live wire and may be dangerous to touch.

The **earth wire** acts as a safety device to prevent electric shock. The earth wire is connected to the casing and if the casing becomes live, because of a fault inside the appliance, a circuit is created through the live and earth wires and large current passes through this circuit. This causes the fuse to melt, thus isolating the appliance.

Look out for

Although in everyday language people might say that a 'fuse had blown', try and avoid this and use words like the 'fuse breaks' or the 'fuse melts', as it is more accurate and shows you have understood the process involved.

Some appliances do not require an earth wire. These are called **double insulated**. The rating plate shows the double insulation symbol.

If you want to be sure you have the correct size of fuse, you can calculate the value of the current in the appliance when it is operating using $I = \frac{P}{V}$. Having calculated the current, the fuse rating used has to be larger than the value of the current; otherwise every time the appliance was switched on the fuse would melt. Using the rating plate above:

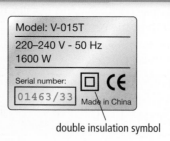

double insulation symbol

$$I = \frac{P}{V}$$

$$I = \frac{1600}{230}$$

$$I = 6{\cdot}95 \text{ A}$$

A fuse with a larger rating than 6·95 A would have to be used, so in this case a 10 A or 13 A fuse would be chosen depending on what is available.

Credit question 1

A lawnmower has a label which shows the following information.

(a) State why this lawnmower has only two wires in the flex. (1)

(b) State the two colours of the wires in the flex. (2)

(c) State the value of the fuse that should be fitted in the plug of this lawnmower. (1)

Happycutter Manufacturing Co

Model HM96–150

230 V a.c. 50 Hz 1500 W

Class II   BEAB approved

Credit question 1 – Answer

(a) The lawnmower is double insulated.

(b) Brown and blue.

(c) 13 A

Note the double insulation symbol on the right side of the rating plate.

If an appliance is double insulated it does not need an earth wire, only live and neutral. The colour of the insulation on the live wire is brown and the colour of the insulation on the neutral wire is blue.

I mark for each correct answer but if you give more than two answers, wrong ones cancel out correct ones

Look out for

There are a number of different fuses that can be fitted in the plugs of electrical appliances, including 3 A, 5 A and 13 A. The one's that are most commonly used are 3 A and 13 A. As a rule of thumb, if the power rating is less than 690 W then a 3 A fuse would be fitted. If the power rating is 690 W or more, then a 13 A fuse would be fitted. The lawnmower in the question has a power rating of 1500 W so a 13 A fuse should be fitted.

Credit question 2

A householder plugs a home entertainment centre, a hi-fi, a games console and an electric fire into a multiway adapter connected to the mains.

The wiring in the electric fire is found to be faulty. The circuit is shown below.

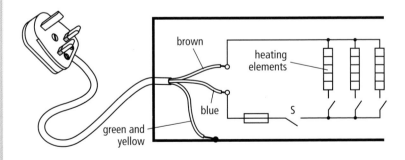

What is the fault in the circuit? (1)

Credit question 2 – Answer

The switch is connected to the neutral/blue wire so that the heating elements would always be live, even if the switch is off.

Look at the circuit carefully. The neutral (blue) and live (brown) wires are the wrong way round and should be swapped or the appliance will still be live even when the switch is open.

Alternating and direct currents

Circuits and circuit diagrams

What you should know at General **and** Credit **level...**

An electric circuit requires a source of electrical energy (often a battery or cell). The **voltage** of the supply is a measure of the energy given to the charges in the circuit. When energy is supplied and there is a complete circuit, then charges will flow around the circuit. The flow of charge per second is called the **current** and the current can be calculated using the equation:

$$I = \frac{Q}{t}$$

Where I is the current in **amperes** (A), Q is the charge in coulombs (C) and t is the time in seconds (s).

Look out for

Although commonly used in everyday speech, 'amps' is not a proper abbreviation of amperes. Although you won't be penalised if you make this mistake at this level, get into the habit of using the proper abbreviation, A.

There are many circuit symbols used to draw circuit diagrams. Some of those you should know are given in the table below:

Component	Circuit Symbol	Component	Circuit Symbol	Component	Circuit Symbol
Cell	—|⊦—	Resistor	—▭—	Fuse	—▭—
Battery	—|⊦---|⊦—	**Diode**	—▷|—	**Capacitor**	—||—
Lamp	—⊗—	Ammeter	—(A)—	Variable Resistor	—▭/—
Switch	—/—	Voltmeter	—(V)—	Two wires connected	—•—

Credit question 1

Two groups of pupils are investigating the following equipment: **ammeter; voltmeter; 12 V d.c. supply; lamp; connecting leads.**

Complete a circuit diagram to show how this equipment can be used to measure the current through, and the voltage across, the lamp.   (3)

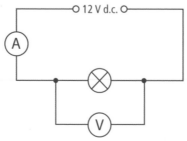

Look out for

The ammeter is always placed in series with the component you want to measure the *current through*; the voltmeter is always placed in parallel with the component you want to measure the *voltage across*. There should be no breaks in the circuit other than between the terminals of the power supply. Use a ruler for the straight lines!

Resistance

Measuring resistance

What you should know at **General** and **Credit** level...

Resistance is the property of a component to oppose current. The larger the resistance, the smaller the current. Resistance can be calculated using the formula:

$R = \dfrac{V}{I}$, where: R is the resistance in ohms (Ω)
V is the potential difference in volts (V)
I is the current in amperes (A).

This is called **Ohm's Law**.

For most components, the resistance remains approximately constant as the current is changed providing the temperature remains the same. There are a number of components that can and do vary their resistance, such as a **variable resistor**; a **light dependent resistor** (LDR); and a **thermistor** (a resistor whose resistance varies with **temperature**). Variable resistors have lots of uses, including the volume control on a hi-fi and the speed control on a Scalextric set.

General question 1

A student sets up circuit 1 to calculate the resistance of resistor R.

(a) Calculate the resistance of resistor R using the meter readings. (2)

(b) The student then sets up circuit 2 to measure the resistance R directly.

Write down the resistance R in ohms obtained from circuit 2. (1)

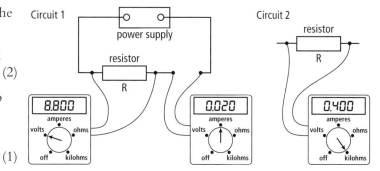

General question 1 – Answer

(a) *The reading on the ammeter is 0·020 amperes and the reading on*

the voltmeter is 8·800 volts.

$R = \dfrac{V}{I}$

$R = \dfrac{8 \cdot 8}{0 \cdot 02} = 440\,\Omega$

(b) *R = 400 Ω*

*Be careful! The meter shown in the diagram is set to read the resistance in kilohms but the question asks for the resistance **in ohms** so you have to convert 0·4 kilohms to ohms by multiplying by 1000.*

In this question make sure you substitute the values from the meters in the correct place in the equation.

Look out for

Although in questions at General level the quantities and units are written out in full, there is no need for you to do this, and you may well find it much easier to use the symbols for both.

Credit question 1

A student uses the circuit below in experiments to investigate how the voltage across different components varies when the current in the components is changed.

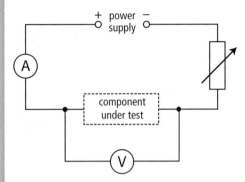

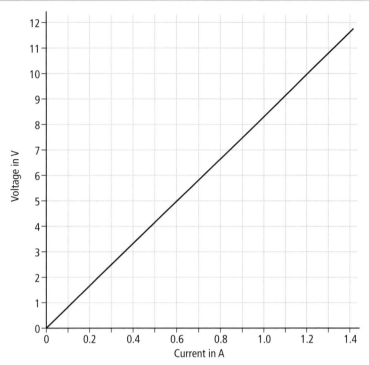

The student places component X in the circuit and carries out an experiment. The graph shows how the voltage across component X varies with current.

(a) Calculate the resistance of component X when the current is 1·2 A.
(You must use an appropriate number of significant figures in your answer.) (2)

(b) Using information from the graph above, explain what happens to the resistance of component X as the current is increased. Justify your answer by calculation or otherwise. (2)

For this question, you first need to find the appropriate voltage from the graph, when the current is 1.2 A, i.e. 10 V.

Credit question 1 – Answer

(a) $R =$

$R =$

$R = 8·3\,\Omega$

(b) *The graph is a straight line which passes through the origin. This tells us that the resistance is constant because as the voltage doubles the current also doubles.*

This can be verified by calculating the resistance using a different point from the graph. For example, when the current is 0·6 A the voltage is 5 V, which gives a resistance of 8·3 Ω.

The question asks you to give an appropriate number of significant figures. Not all questions will give you this clue. The answer you get on the calculator for this question is 8·333333333 but do not leave that as your final answer. The current and voltage are each given to two significant figures, so give your answer to two significant figures i.e. 8·3 Ω. Examiners will not penalise you if you get the number of significant figures in an answer slightly wrong. In this question you could have given your answer as 8, 8·3, 8·33 or 8·333 but writing 8·3333 would be considered too many and you would lose a ½ mark.

Look out for

Remember the general rule on significant figures is you are allowed between one less and two more significant figures than the smallest number in the data. Do not quote everything to 2 decimal places; it may work when dealing with the question above but a number such as 1098·12, whilst only having two decimal places, is six significant figures. Never use the recurring sign (i.e. a dot above the last number) – this is telling the examiner that you know the answer is accurate to an infinite number of figures, which is impossible!

Credit question 2

Using the circuit in the previous question, the student replaces component X with component Y, repeats the experiment and obtains this graph.

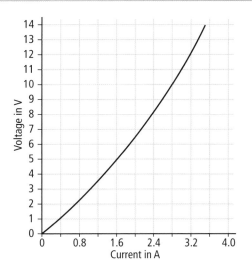

The student concludes that the resistance of component Y is not constant. Why is the student correct in coming to this conclusion?

(1)

Credit question 2 – Answer

The gradient (slope) of the line is increasing as the current increases, so the resistance must be increasing.

If you weren't sure about what was happening to the resistance in this part, then the fact that the resistance is increasing could be confirmed by calculating the resistance at two different points on the graph.

Look out for

This is the type of graph you would expect to get when the temperature was increasing; for example in a filament lamp as the voltage across it is increased. For many conductors, as the temperature increases, so does the resistance.

Energy and power

What you should know at **General** and **Credit** **level...**

In many resistive circuits, electrical energy is converted to heat energy: for example, an electrical heater or a filament lamp. The rate at which energy is transformed is known as the **power** (we often talk about the power dissipated by a component). Power is measured in watts. One watt is equal to one **joule** per second.

The power can be calculated in a number of ways:

1 $P = \dfrac{E}{t}$, where:

P is the power in watts (W), E the energy in joules (J) and t the time in seconds (s).

continued

27

What you should know at General and Credit level – continued

2 $P = IV$, where:
  P is the power in watts (W), I the current in amperes (A) and V the voltage in volts (V).

3 $P = I^2R$, where:
  P is the power in watts (W), I the current in amperes (A) and R the resistance in ohms (Ω).

It is also useful to know the following equation:

$P = \dfrac{V^2}{R}$, where

P is the power in watts (W), V is the voltage in volts (V) and R the resistance in ohms (Ω).

Look out for

We can show that the relationships in 2 and 3 are equivalent as follows:

Starting with $P = IV$ and $V = IR$

We substitute IR in place of V in the first equation to give:

$P = IIR$ which is

$P = I^2R$

General question 1

A hairdryer and its rating plate are shown below.

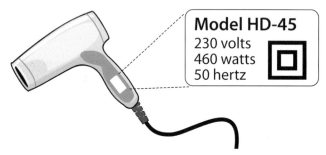

Model HD-45
230 volts
460 watts
50 hertz

Calculate the current in the hairdryer when it is operating.   (2)

Look out for

The information needed to do this question is given in the rating plate on the appliance. Some of the information given isn't required, for example the frequency. If you're not sure what bits of information to use, write them down first and then think of or look for the equation that links them. You don't have a formula that includes power and frequency so that will tell you that the frequency is not required.

General question 1 – Answer

$P = IV$
$460 = I \times 230$
$I = \dfrac{460}{230}$
$I = 2\ A$

Credit question 1

A 25 W lamp is designed to be used with mains voltage.
Calculate the resistance of the lamp.

Credit question 1 – Answer

$V = 230\ V,\ P = 25\ W$
$P = \dfrac{V^2}{R}$
$25 = \dfrac{230^2}{R}$

$R = \frac{230^2}{25}$

$R = 2116 \ \Omega$

This question can be tackled in a number of ways. The first thing to identify is that the lamp operates on mains voltage, which is 230 V. You could then calculate the current using $P = IV$ and then find the resistance using Ohm's Law, $R = \frac{V}{I}$.

Look out for

Being able to rearrange formulae is an important skill but when carrying out calculations like the one in this example, it is advisable to write down the formula, substitute the numbers and then rearrange to find the quantity you are looking for. If you then make a mistake, the examiner will treat it as an arithmetic mistake and you will only lose ½ mark, whereas if you rearrange the formula incorrectly the maximum you will gain is ½ mark for the original formula, if it is correct.

Lamps and heaters

What you should know at **General** and **Credit** level...

Traditional **filament lamps** transform electrical energy into heat and light. In fact around 90% of the energy transformed by the filament is given off as heat and only around 10% is given off as light.

The filament is a resistance wire. This is where the energy is transformed.

By contrast, in a **discharge tube**, the electrical energy is converted to light in the gas inside the tube. Discharge tubes are much more efficient than filament lamps as they generate more light for the same amount of electrical energy as an equivalent filament light. (This also means they generate much less heat.)

The energy transformation takes place in the gas inside the discharge tube.

In a heater, the energy transformation takes place in the resistance wire (often called the heating element).

General question 1

Conventional filament lamps are now being replaced by discharge tubes.

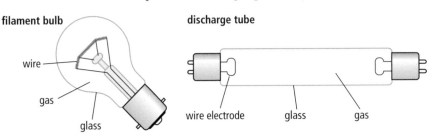

State where the energy transformation occurs in:

(a) the filament lamp; (1)

(b) the discharge tube (1)

(c) State why discharge tubes are replacing conventional filament lamps. (1)

General question 1 – Answer

(a) *In the wire*

(b) *In the gas*

(c) *Discharge tubes are more efficient **or** cost less to run in the long term **or** more light for same power or converse **or** save energy **or** lasts longer **or** produce less heat (and more light)*

Watch out for 'loose' answers such as 'eco friendly', 'cheaper' (which is incorrect as discharge tubes can cost more to buy) and 'no heat from discharge tubes' (they do give off some heat). These answers would have gained you 0 marks.

Useful circuits

Series and parallel circuits

What you should know at **General** and **Credit** **level...**

In circuits, components can be connected in series or parallel. In a **series circuit**, the components are connected one after the other to form a single complete path for the charge to flow. In a **parallel circuit** there are different paths or 'branches' for the charge to flow.

In a series circuit:

▶ the current is the same at all points;

▶ the sum of the voltages across the components is equal to the supply voltage.

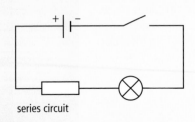

series circuit

continued

What you should know at General and Credit level – continued

In a parallel circuit:

▸ the sum of the currents in parallel branches is equal to the current from the supply;

▸ the voltage across parallel branches is the same for each branch.

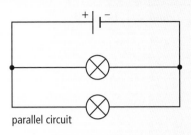

parallel circuit

Switches are used in circuits to control components. More than one switch can be used in a circuit and switches can be connected in series or in parallel.

General question 1

Two identical lamps are connected to a 6·0 volt battery as shown in circuit 1.

The battery supplies a current of 0·40 ampere to the circuit.

Complete the table to show the current in each lamp and the voltage across each lamp. (2)

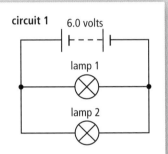

General question 1 – Answer

	Lamp 1	Lamp 2
Current (amperes)	0·20	0·20
Voltage (volts)	6·0	6·0

The current from the battery has two paths it can take and the two lamps are identical, so the current in each lamp will be half the current from the battery. The lamps are connected in parallel with the battery so the voltage across each will be the same as the battery.

General question 2

Three identical lamps are shown in Circuit 1.

The battery has a voltage of 12 volts and supplies a current of 0·9 ampere to the circuit.

Complete the table below to show the current in each lamp and the voltage across each lamp. (2)

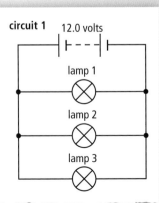

General question 2 – Answer

	Lamp 1	Lamp 2	Lamp 3
Current (amperes)	0·3	0·3	0·3
Voltage (volts)	12	12	12

The circuit has three branches in parallel with identical lamps, so the current in each branch will be one third of the current from the battery. The voltage across each lamp will be the same as the supply voltage since the lamps are connected in parallel.

General question 3

The three lamps and battery are now connected as shown in Circuit 2 below. The current from the battery is now 0·1 ampere.

Complete the table below to show the current in each lamp and the voltage across each lamp.   (2)

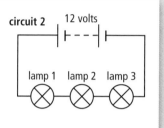

General question 3 – Answer

	Lamp 1	Lamp 2	Lamp 3
Current (amperes)	0·1	0·1	0·1
Voltage (volts)	4	4	4

Circuit 2 is a series circuit, so the current is the same at all points as there is only one path. The voltage splits across each component and the total adds up to the supply voltage.

Look out for

In descriptive questions about circuits never talk about 'voltage through' a component. This is incorrect and you will get 0 marks. Always use the term 'voltage across'.

Resistors in series and parallel

What you should know at  **General** and **Credit** level...

The combined resistance of resistors connected **in series** can be found using the relationship:

$R_T = R_1 + R_2 + \ldots$

The combined resistance of resistors connected **in parallel** can be found using the relationship:

$\frac{1}{R_T} = \frac{1}{R_1} + \frac{1}{R_2} + \ldots$

Look out for

The combined resistance of resistors in parallel is always smaller than the smallest individual resistor in the parallel circuit.

Credit question 1

A student sets up a circuit to measure the resistance of a number of resistors.

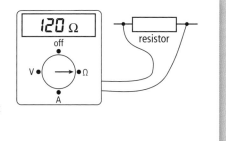

The resistances of the other three resistors are 180 Ω, 220 Ω, and 360 Ω. The student connects all four resistors in series.

Calculate the total resistance. (2)

Credit question 1 – Answer

$$R_T = R_1 + R_2 + R_3 + R_4$$
$$R_T = 120 + 180 + 220 + 360$$
$$R_T = 880 \ \Omega$$

Questions on resistors in series don't get asked too often in this way. The only really tricky part with the question was spotting that one of the resistances is given on the meter.

Credit question 2

A 5A circuit breaker is used in a household lighting circuit which has three 60 W lamps as shown. The resistance of each lamp is 882 Ω

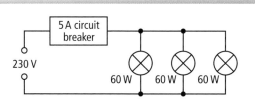

Calculate the combined resistance of the three lamps in this circuit. (2)

Credit question 2 – Answer

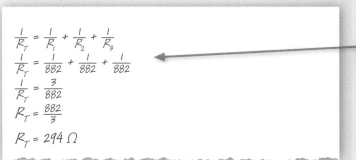

$$\frac{1}{R_T} = \frac{1}{R_1} + \frac{1}{R_2} + \frac{1}{R_3}$$
$$\frac{1}{R_T} = \frac{1}{882} + \frac{1}{882} + \frac{1}{882}$$
$$\frac{1}{R_T} = \frac{3}{882}$$
$$R_T = \frac{882}{3}$$
$$R_T = 294 \ \Omega$$

Be very careful how you set your answers out when working out total resistance in parallel. One of the most common mistakes is to write the formula as $R_T = \frac{1}{R_1} + \frac{1}{R_2} + ...$ You will score 0 marks if you do this.
Also don't be tempted to write things like $\frac{3}{882} = \frac{882}{3} = 294 \ \Omega$. This is known as 'bad form' and you will lose marks because of it.

Look out for

When all the resistors in parallel have the same value then there is an easy shortcut you can use. In the above example each resistor has a resistance of 882 Ω and there are three of them so the combined resistance is $\frac{1}{3}$ of 882. If there had only been two identical resistors in parallel, then the combined resistance would have been ½ of the individual resistances.

Note that this shortcut only works when the individual resistances are equal; if they are different you have to use the method shown in the example.

Credit question 3

The settings of a variable speed control use different combinations of identical resistors as shown.

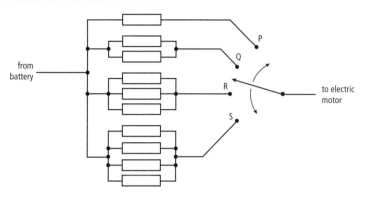

from battery

P
Q
R
S

to electric motor

(a) To which position should the variable speed control be set to achieve maximum speed? (1)

(b) Justify your answer. (1)

Credit question 3 – Answer

(a) *Position S*

(b) *This position gives maximum current **or** this position gives least/minimum resistance **(must be 'least' or equivalent, not 'less')** **or** this position gives greatest/maximum voltage across the fan.*

Position S has the most identical resistors connected in parallel (four), so the combined resistance will be ¼ of the individual resistance. The least resistance will give the greatest current.

*Your answer must say **least** (or an equivalent such as **smallest**, or **lowest**), not **less**.*

Fault finding

What you should know at **General** and **Credit** **level...**

A simple fault finder for circuits, called a **continuity tester**, consisting of a battery, a lamp, two probes and connecting wires, can be used to detect open circuits.

An **open circuit** is where there is a break in the circuit where there shouldn't be one: for example, a broken lamp. When the circuit is tested with the continuity tester, the bulb would not light if there was a break.

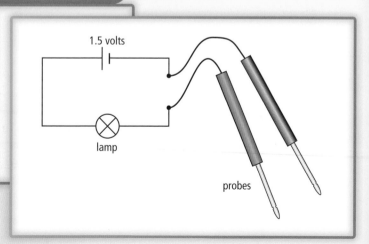

1.5 volts

lamp

probes

General question 1

Party lights consist of 16 identical light bulbs connected in series. They operate from a 24 volt power supply. The current in the circuit is 1·25 amperes.

A fault occurs in the circuit and a continuity tester is needed to find the fault. The circuit diagram for the continuity tester is shown.

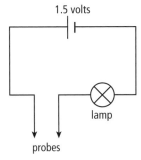

(a) Describe how the continuity tester could be tested to make sure that it is working. (1)

(b) The continuity tester is found to be faulty. State one possible reason why it is not working. (1)

General question 1 – Answer

(a) *Connect the probes together and see if the lamp lights.*

(b) *Battery flat /voltage too low **or** broken/loose wire **or** open circuit*

*   *or lamp faulty/broken **or** lamp short circuited*

There are lots of possible reasons why the continuity tester may not be working. Some answers that wouldn't be accepted are 'lamp blown' which is another example of loose language which shouldn't be used and 'short circuit' on its own. You need to explain what was short circuited, for example, the lamp.

Look out for

You should always test a continuity tester in this way before you use it, in case the tester itself is faulty.

Circuit faults can also be detected by using an ohmmeter. There are two types of fault that can be detected: an **open circuit** or a **short circuit**. For an open circuit the ohmmeter reading would 'go off the scale' because there is a break in the circuit. For a short circuit the ohmmeter would give a much smaller reading of resistance than expected.

Behind the wall

The mains supply

All household appliances are connected in parallel via the household mains wiring. **Lighting circuits** are used for the fixed lighting in your house; they use a 5 A fuse or circuit breaker and have thinner wiring than that used for the mains sockets.

The power sockets are supplied using a **ring main circuit**. As the name suggests, this circuit has a complete loop, which means there are two paths for the current, so the circuit can use thinner wire than would be required if it was a normal parallel circuit. The wiring in a ring main circuit is thicker than that of a lighting circuit because it may need to carry a greater current. Ring main circuits use a 30 A fuse or **circuit breaker**. Appliances such as cookers or immersion heaters have their own circuits because of the high power rating of the appliances, and therefore the high currents in them.

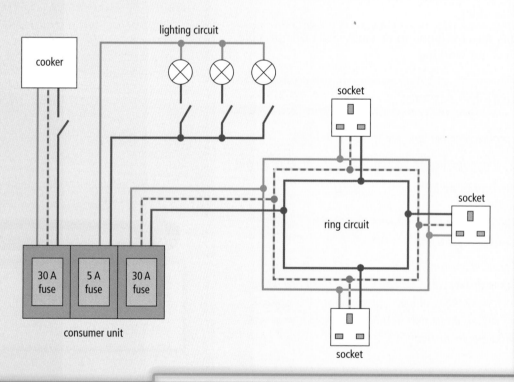

The household wiring is protected either by fuses in the consumer unit or, more commonly these days, by circuit breakers. Circuit breakers are automatic switches which operate quickly to stop the current if it gets too big. They tend to be faster acting than traditional fuses and they are easy to reset, whereas a fuse in the consumer unit has to have the fuse wire replaced if it melts. The mains fuses or circuit breakers protect the mains wiring.

The electricity meter you have in your home measures the energy used in kilowatt hours (kWh). One **kilowatt hour** is equivalent to 3 600 000 joules (1000 × 60 × 60).

Credit question 1

The diagram shows three household circuits, connected to a consumer unit.

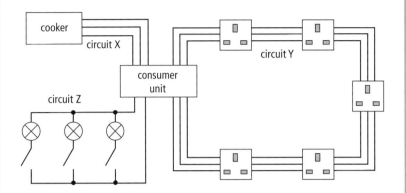

(a) Which circuit is a ring circuit? (1)

(b) Give two advantages of using a ring circuit. (2)

(c) State and explain one difference between a lighting circuit and a ring circuit. (2)

(d) Why does a cooker need a separate circuit? (1)

(e) One heating element of the cooker has a power rating of 2·2 kW. Calculate how many joules of energy are transferred by this element in 2 hours. (2)

Look out for

Remember the Credit paper contains a table of SI prefixes and multiplication factors on the data sheet.

All three wires, Live, Neutral and Earth make a loop or ring from the consumer unit.

Credit question 1 – Answer

(a) *circuit Y*

(b) *Cheaper **or** less current in each branch **or** thinner wire **or** convenient for adding extra sockets **or** 2 paths for current **or** less voltage drop or less heat*

(c) *The lighting circuit is a simple parallel circuit because the current in it is lower; the lighting circuit also has thinner wire for the same reason.*

(d) *The cooker has a high power rating so requires a larger current, so it needs a separate circuit.*

(e) $E = Pt$

$E = 2.2 \times 1000 \times 2 \times 60 \times 60$

$E = 15\,840\,000$ J

The following answers would not be accepted: less wire **or** less current per ring **or** any comparison with series circuit **or** safer

There are in fact a number of possible answers including the different fuse ratings but the question specifically asks for one difference, so do not be tempted to give more, in case wrong answers cancel out correct ones.

In this question the power rating is given in kilojoules, so has to be converted to joules; and the time is given in hours, so must be converted to seconds.

Movement from electricity

Electric motors

Electric motors operate by using the fact that when a current passes through a wire, a magnetic field is produced around the wire.

When the wires are coiled and carry a current then this produces an electromagnet. The resulting magnetic field interacts with the magnetic field from either permanent magnets, in a model motor, or field coils in a commercial motor.

The direction of the force produced on a current carrying conductor in a magnetic field depends upon the direction of the magnetic field and the direction of the current.

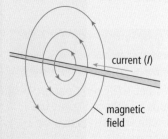

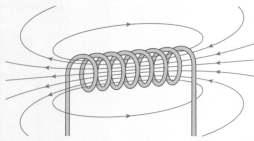

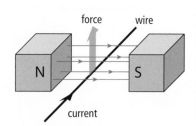

In a simple electric motor, one side of the coil is forced up and one side forced down, as shown in the diagram below.

The **brushes** provide electrical contact to the coil and the **commutator** reverses the direction of the current in the coil every half turn to ensure that the current always travels in the same direction through the field. This, in turn, ensures that the force on the coil is always in the same direction and the coil keeps spinning.

A commercial electric motor has a number of differences to the model electric motor you may have used in the laboratory.

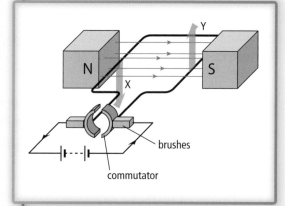

The commercial motor uses carbon brushes because they provide low friction and good electrical contact but do not wear away the commutator. The commutator is split into multiple sections, two for each **rotor coil** since there are several rotor coils rather than one. Having several rotor coils means that there will always be a coil in the correct plane of the magnetic field, this means the motor can be 'self-starting' and gives a much smoother rotation. **Field coils** are used rather than permanent magnets; these field coils can give a stronger magnetic field for the same size of magnet.

The magnetic effect of a current isn't just used in electric motors. Devices such as relays, electric bells and **solenoids** also use this effect. For example, a relay is an electromagnetically operated switch. It contains a coil and when a current passes through the coil a magnetic field is created which attracts a contact and closes the switch.

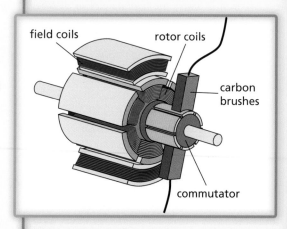

Credit question 1

A simple d.c. motor is shown below:

The coil WXYZ rotates in a clockwise direction.

State **two** changes that could be made to make the coil rotate in the opposite direction.

(2)

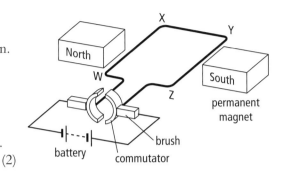

Reversing or swapping about the poles on the magnets will reverse the direction of the magnetic field. Reversing the battery connections will reverse the direction of the current in the coil. Note that if you wanted to reverse the direction of rotation you would actually only make one of these changes. Making both would result in the coil spinning in the same direction as before.

Credit question 1 – Answer

Reverse the poles on the magnets and reverse the battery connections.

Credit question 2

Part of a commercial motor is shown below

In the commercial electric motor, state why

(a) more than one rotating coil is used

(1)

(b) field coils rather than permanent magnets are used.

(1)

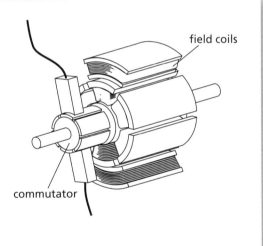

Credit question 2 – Answer

(a) *Having more than one rotating coil gives a smoother rotation **or** means the motor will start more easily **or** can produce a greater rotating force.*

Note that only one reason is required for each of parts (a) and (b).

(b) *Using field coils is easier to control **or** the field can be shaped more easily **or** they can give a stronger magnetic field for the same size of permanent magnets.*

The use of thermometers

The use of thermometers in measuring body temperature

What you should know at **General** **level...**

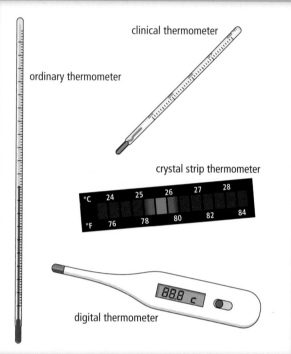

clinical thermometer

ordinary thermometer

crystal strip thermometer

digital thermometer

Thermometers are used to measure temperature. Temperature tells us how hot or cold a body is and is measured in °Celsius.

Body temperature is important in the diagnosis of illness. Normal body temperature is 37°C and when it increases it can indicate fever or coldness when it decreases.

A thermometer needs some measurable physical quantity which changes with temperature typically the expansion of a liquid. An ordinary liquid-in-glass thermometer has a thin bore within a glass tube which contains a liquid in the bulb. When heated, the liquid expands and rises up the tube. When the liquid cools, it contracts and falls down the tube.

A clinical glass thermometer is similar to an ordinary liquid-in-glass thermometer, but differs in two ways:

▶ it has a smaller range of temperature scale

▶ it has a small kink or constriction in the tube to prevent the liquid falling back when it is removed from the patient to allow the temperature to be read.

To measure body temperature:

▶ place the thermometer in the mouth. Wait a suitable time until it reaches the body temperature. Remove the thermometer and note the reading on the scale.

▶ remember to remove the thermometer from the mouth – it is very difficult to get a reading whilst still in the mouth!

Look out for

Make sure you know the difference between **temperature** and **heat** – heat is a measure of an object's energy measured in joules which is related to its temperature.

General question 1

Several types of thermometers are shown.

(a) What is the purpose of a thermometer?

(b) Clinical thermometers are designed for medical use. Other thermometers (sometimes called 'ordinary thermometers') are made for general laboratory use.

Describe **two** important differences between a clinical thermometer and an 'ordinary thermometer'.

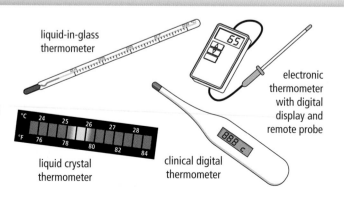

liquid-in-glass thermometer

electronic thermometer with digital display and remote probe

liquid crystal thermometer

clinical digital thermometer

General question 1 – Answer

(a) *A thermometer is used to measure the temperature of an object.*

(b) *Difference 1: Smaller range of temperature*

   Difference 2: It has a kink in the tube.

Note that a thermometer does not measure heat. Temperature is an indication of how hot or cold something is.

Not simply that it is smaller.

Using sound

The stethoscope

What you should know at General **level...**

Sound is produced by the vibrations of particles. A solid, liquid or a gas is needed for the transmission of sound. Therefore a vacuum will not allow the transmission of sound because it does not contain any particles.

In medicine sound from the body can give information about blood pressure, lung and heart function using a stethoscope to listen to sound frequencies and rhythms. The stethoscope has two bells to detect different frequencies of sound. Sound travels along tubes from the bells to the earpieces.

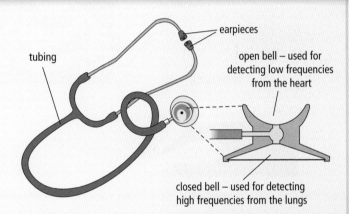

earpieces

tubing

open bell – used for detecting low frequencies from the heart

closed bell – used for detecting high frequencies from the lungs

General question 1

A doctor uses a stethoscope like the one shown above to listen to the sounds of a patient's heart.

(a) Explain how the stethoscope acts as a hearing aid for the doctor. Your explanation must give the purpose of each of the parts labelled in the diagram. (3)

(b) Why is it important that the bell makes firm contact with the patient's body? (1)

General question 1 – Answer

(a) *Bell: detects sound from the body.*

   Rubber tubing: allows sound to travel from the bell to the earpieces.

   Earpieces: transmit sound to inside the ear for the doctor to listen to.

(b) *To maximise the sound from the body or To minimise sounds from outside the body*

Note that three marks indicates you must make three statements.

Ultrasound and ultrasonic scanning

What you should know at **General** and **Credit** level...

High frequency vibrations beyond the frequency range of human hearing (above 20 000 Hz) are called **ultrasounds**.

An example of the use of ultrasound in medicine is to image unborn babies. Images of internal organs and tissue are created by ultrasound by sending waves into the body using a probe. When they pass from one type of tissue to another, some waves are reflected back to a receiver in the probe. A computer is used to analyse the times of the reflected ultrasounds to produce an image of the inside of the body. Ultrasounds are regularly used to monitor the development of unborn babies and are much safer than X-ray scanning which may be harmful.

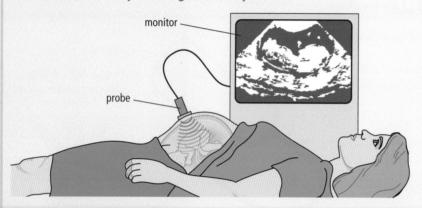

monitor

probe

Look out for

Be careful when defining ultrasound. You must include the term 'frequency'. Simply saying that ultrasounds are beyond the range of human hearing will not get any marks.

Credit question 1

Ultrasound is used by doctors for treatment and diagnosis. Pulses of ultrasound are used to produce local heating of muscle deep inside the body. This heating effect can help relieve pain in the muscles.

(a) What is meant by ultrasound?

(b) Calculate the time for a pulse of ultrasound to travel through 2 cm of muscle. (You will find the data you need on p 2 of the Data Sheet.)

Credit question 1 – Answer

(a) *Sounds which have frequencies greater than 20 000 Hz* **or** *sounds above the frequency range of humans.*

(b) $V = \dfrac{d}{t}$

$1500 = \dfrac{0.02}{t}$

$t = \dfrac{0.02}{1500} = 1.333 \times 10^{-5} s$

$= 1.3 \times 10^{-5} s$

Selection of this value gains one mark.

Remember to convert cm to m.

Remember to round your answer to a sensible number of significant figures.

Credit question 2

Ultrasound is used to build up images of an unborn baby.

(a) Explain how ultrasound is used to build up such images.

(b Why is ultrasound safer than **X-rays** for this sort of medical application?

Credit question 2 – Answer

(a) *Ultrasound waves are sent into the body using a probe. When they pass*

from one tissue to another some waves are reflected to the probe.

A computer analyses the reflected sounds and produces an image.

(b) *X-rays can damage cells but ultrasound is safe to use.*

Noise pollution

What you should know at General level...

Learn some values of sound levels. Note that the threshold of hearing is 0 dB.

Noise pollution (or environmental noise) is unpleasant human, animal or machine-created sound that interrupts normal life. Common sources are noises from road and air traffic, pop concerts and noisy neighbours. It is important to measure and monitor noise levels as prolonged exposure can cause damage to the ears resulting in deafness. Sound levels are measured in **decibels (dB)**, and damage can be caused by noises over 90 dB. People regularly exposed to levels above this should wear ear protectors to absorb the sound energy.

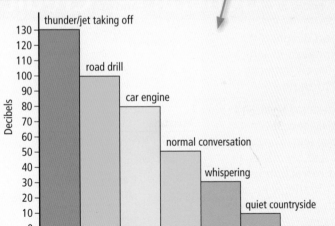

General question 1

The table below gives sound levels from different sources of sound.

Source of sound	Sound level (dB)
1 metre from a disco loudspeaker	120
5 meters from a pneumatic drill	100
beside a busy motorway	90
inside the cab of a tractor	90
inside a busy supermarket	70
inside a busy office	60
normal conversation	50

When one source of sound is twice as loud as another, the sound level increases by 10 dB.

(a) Which of the above sources of sound is twice as loud as the level inside a busy office?

(b) When working in very noisy surroundings, what precaution should a person take to guard against hearing damage?

General question 1 – Answer

(a) *Inside a busy supermarket. The busy office is 60 dB, therefore an increase of 10 dB will make the sound of the busy supermarket twice as loud.*

(b) *He/she should use ear protectors.*

Light and sight

Refraction

Light was introduced in Chapter 1. Revise the meanings of the terms incidence, the normal and the law of reflection. Refer back to p 8 for advice on measuring angles etc.

What you should know at **General** **and** **Credit** **level...**

Refraction of light is the change in speed of light when it travels from one material to another.

The incident ray makes an angle of incidence with the normal and likewise the refracted ray makes an **angle of refraction** with the normal. You should be able to identify these angles from ray diagrams.

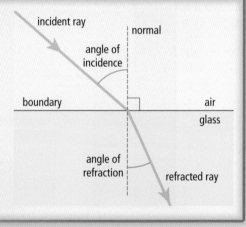

⚠ Look out for

Be careful – refraction is not the change in direction of the light. When light travels along the perpendicular, or normal, to the surface it does not change direction but it is still refracted. Learn the spelling of refraction and do not confuse it with reflection.

Lenses

Lenses are shaped pieces of glass or plastic which use the principle of refraction to alter the paths of light. These are necessary to correct eye defects. The two main shapes of lenses are **convex** and **concave**:

Note that thick lenses have shorter focal lengths compared to thin lenses. The thick lenses are able to bend the light more so are more powerful.

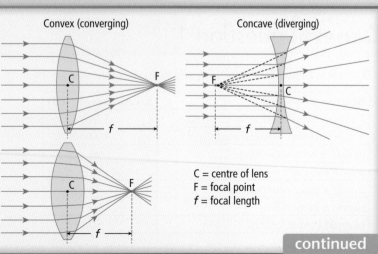

C = centre of lens
F = focal point
f = focal length

continued

What you should know at General and Credit level – continued

A simple experiment can be used to determine the focal length of a convex lens:

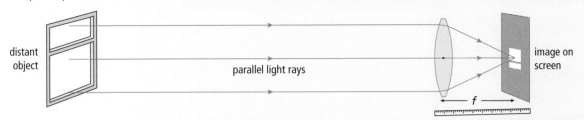

You need a lens, a distant light source, a screen, and a ruler.

▶ Use the lens to project an image of the distant source (such as a window) onto the screen.

▶ Adjust the position of the lens until the image is sharply focused.

▶ Use a ruler to measure the distance from the lens to the screen – this is the **focal length**.

General question 1

A class investigates the effects of the following shapes of glass on rays of white light.

The teacher sets up three experiments, covering the glass shape with card.

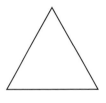

The paths of the light rays entering and leaving the different shapes of glass are shown.

For each of the three experiments, draw the **shape** and **position** of the glass block that was used.

(a)   (b)   (c)

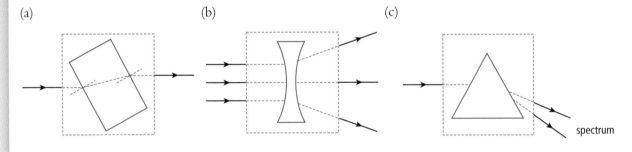

Credit question 1

In a physics laboratory, a student wants to find the focal length of a convex lens. The student is given a sheet of white paper, a metre stick and a lens.

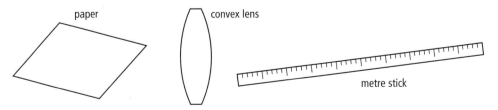

(a) Explain how the student could measure the focal length of the lens using this equipment.

(b) Refraction of light occurs in lenses. What is meant by the term refraction?

Credit question 1 – Answer

(a) <u>Hold lens between a distant object and the paper. Use the lens to focus/get sharp/clear image. Measure distance from lens to paper with the ruler.</u>

(b) <u>Refraction is when speed/wavelength changes when light travels from one material to another.</u>

This is the official SQA answer so learn it.

Credit question 2

A student wears spectacles for reading and decides to measure the **focal length** of one of her spectacle lenses. She sets up a screen, a brightly lit bulb and a metre stick as shown in the diagram below.

She moves the screen until a clear image of the bulb is obtained and then measures the distance XY between the lens and the screen.

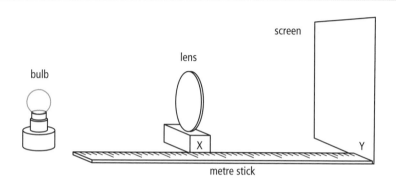

(a) Give a reason why the distance XY is not equal to the focal length of the lens.

(b) State the change she should make in carrying out her experiment so that she measures the focal length of the lens.

Credit question 2 – Answer

(a) <u>The incoming light from the bulb does not provide parallel rays.</u>

(b) <u>She should use light from a distant source (which produces parallel rays) such as a window or move bulb much further away from the lens.</u>

Image formation

What you should know at **General** **level...**

Visible light enters the eye through the pupil (an aperture or opening) having passed through the cornea where most of the light is refracted. The lens changes shape to focus on near and distant objects. (This process is called **accommodation**). At the retina light energy is converted into electrical signals which are carried to the brain by the optic nerve.

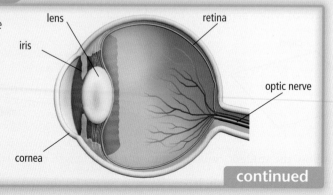

continued

What you should know at General level – continued

The image formation on the retina of the eye is upside down and laterally inverted as shown in the diagram. Notice how a ray of light from the head reaches the retina below a ray of light coming from the feet, so the images are inverted on their way to the retina at the back of the eye. The same happens with rays of light coming from the sides to produce sideways inversion.

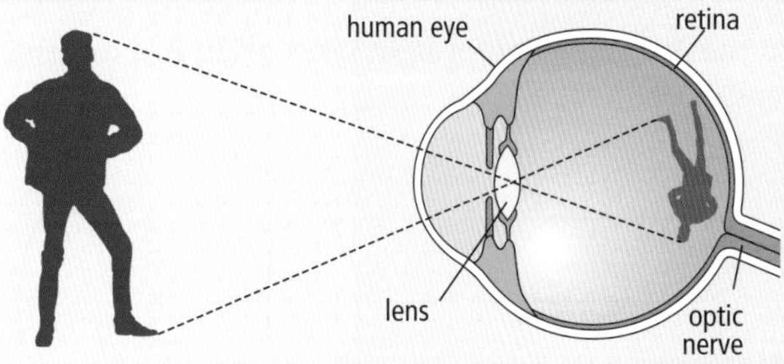

General question 1

A student is looking at a sheet of paper which has the letter F drawn on it as shown here. **F**

Which is the correct image of the letter formed on the student's retina?

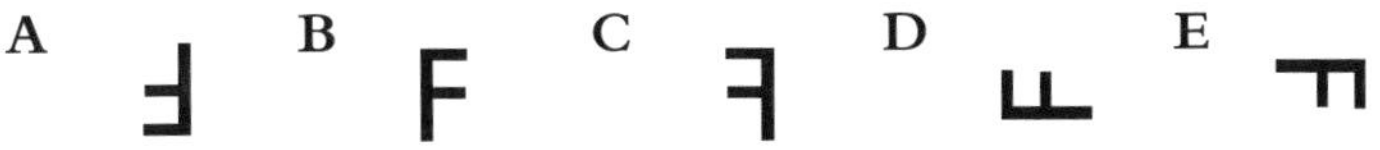

General question 1 – Answer

A

A is the only one which is upside down and back to front.

What you should know at Credit level...

To explain how the lens of the eye forms images on the retina, you need to know how light passes through a lens. There are two important rules to remember when showing how an image is formed with a lens:

1 A ray of light from the object which passes through the centre of the lens (C) goes straight through.

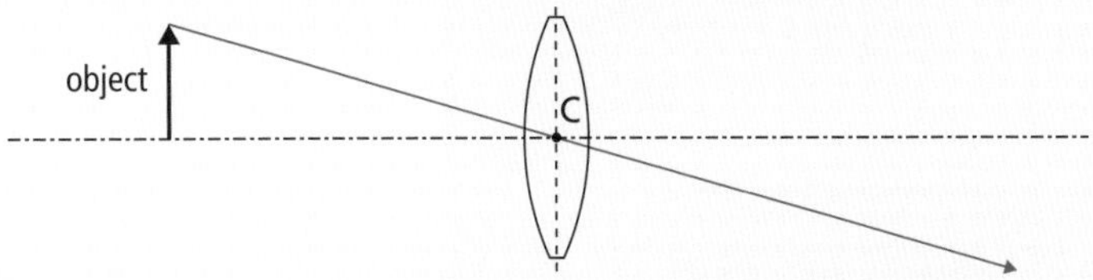

2 A ray of light which enters the lens travelling horizontally is refracted through the focal point (F).

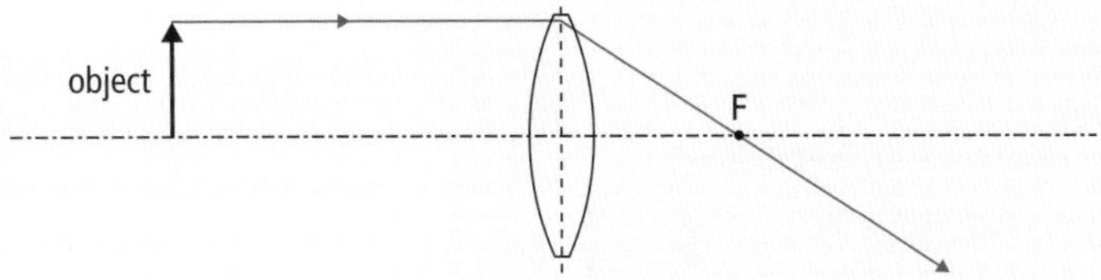

These two rays meet at the image (**I**).

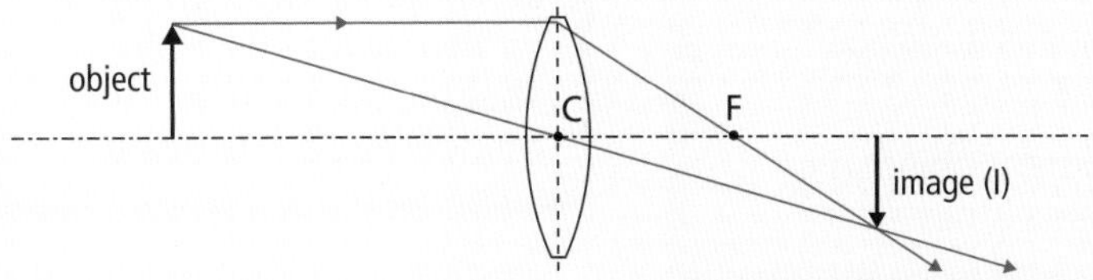

These principles are applied to the lens in the eye to locate images when focussing on near and distant objects as shown-

The focal length of a lens is used to calculate the power of a lens. You must be able to carry out calculations which involve the focal length *f* and power *P* of lenses.

$$P = \frac{1}{f} \text{ or } f = \frac{1}{P}$$

Where *P* is power measured in **dioptres** (D), and *f* is the **focal length** measured in metres (m).

Look out for

Be careful – do not confuse optical power with electrical or mechanical power which are measured in watts. Remember the focal length must be in metres – this is a very common mistake.

Credit question 1

In the eye, refraction of light takes place at the cornea and the eye lens.

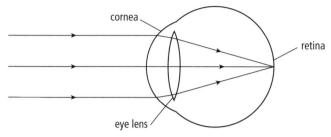

More of the refraction takes place at the cornea as shown above. To show how the eye forms an image, a student uses two identical lenses and a screen to make a model eye. Three parallel rays of light are directed towards the lenses and are focussed on the screen as shown.

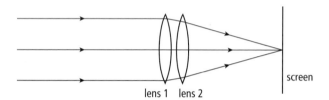

State one change that could be made to the lens system to represent more correctly the eye.

Credit question 1 – Answer

Lens 1 should be thicker to represent the cornea where more of the refraction occurs.

Correction of eye defects

What you should know at General **level...**

Long-sightedness and short-sightedness are common eye defects which can be corrected using lenses.

Long sight: the focal point of light rays from a nearby object is behind the retina so the image formed on the retina is not clear. (Note that the rays do not focus behind the retina.)

Short sight: the focal point of light rays from a distant object is in front of the retina so the image formed on the retina is not clear.

Look out for

Be careful – do not state that people cannot see objects at a distance for short sight (or nearby for long sight). They are not blind and simply cannot see objects clearly.

The diagrams below show the ray diagrams for normal eyes and for short sighted and long sighted eyes. Concave and convex lenses are used to correct long-sightedness and short-sightedness. Laser surgery can also be used to correct short sight by reducing the curvature of the cornea in the eye. This makes the focal length longer.

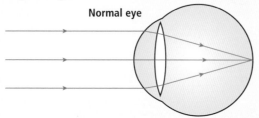

Normal eye

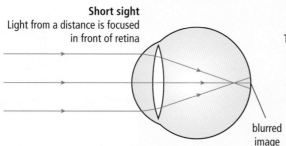

Short sight
Light from a distance is focused in front of retina

blurred image

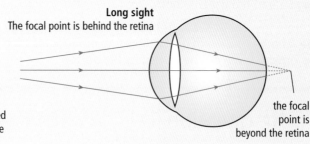

Long sight
The focal point is behind the retina

the focal point is beyond the retina

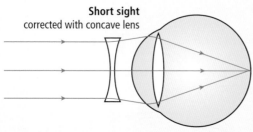

Short sight
corrected with concave lens

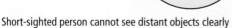

Short-sighted person cannot see distant objects clearly

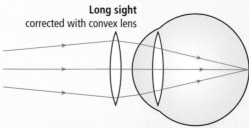

Long sight
corrected with convex lens

Long-sighted person cannot see close-up objects clearly.

Credit question 1

A short–sighted person needs a lens of power $-4\cdot0$ D to correct their short sight.

(a) What is the shape of this lens?

(b) Calculate the focal length of the lens.

Credit question 1 – Answer

(a) _Concave (because of the negative focal length)_

(b) $P = \dfrac{1}{f}$

$4 = \dfrac{1}{f}$

$f = \dfrac{1}{4} = 0\cdot25\,m$

Remember the focal length **must** be in metres.

The use of fibre optics in medicine

What you should know at General and Credit **level...**

Optical fibres are thin strands of glass along which light is able to travel by the process of **total internal reflection**. (Go back to Chapter 1 to revise this topic).

An endoscope (or fibroscope) is an instrument which contains optical fibres and is often used by doctors to examine the inside of the body e.g. the stomach. The light source should emit very little heat which could irritate internal organs.

The **light guide** is the bundle of optical fibres used to illuminate the internal organs. Light from the source travels through it into the patient's body.

Light from the stomach reflects back up through the **image guide** (another bundle of optical fibres) to produce an image of the organs which can be viewed by the doctor.

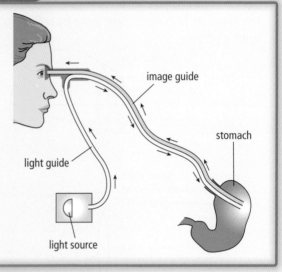

Credit question 1

A health physicist is developing a system for measuring temperatures inside the body. A thermocouple is inserted through a tube beside the optical fibres of an endoscope. The endoscope allows the doctor to see where the thermocouple is being positioned. The endoscope consists of two fibre bundles and a 'cold light' source.

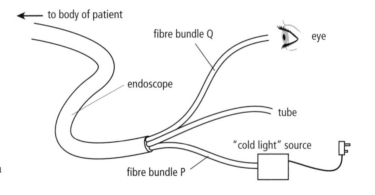

(a) Explain the purpose of each of the two bundles of fibres in the endoscope.

(b) What is meant by 'cold light' source?

(c) Explain whether a filament lamp or a discharge lamp would be more suitable for the light source of the endoscope.

Credit question 1 – Answer

(a) *Fibre bundle P: Light travels from the cold light source into the body of the patient.*

Fibre bundle Q:

Light travels along Q from the patient to the doctor.

(b) *A cold light source emits light only and no heat.*

(c) *Discharge lamp. This produces less heat than a filament light.*

Using the electromagnetic spectrum

Use of lasers

What you should know at General **level...**

A laser is a device used to produce a very concentrated beam of light. In medicine, lasers can be used to:

▶ cut or vaporise tissue. The laser beam is directed into the body using an endoscope and small tumours (cancer cells) can be removed.

▶ seal damaged blood vessels which stops unwanted bleeding

▶ correct short sight by removing part of the cornea.

Learn at least one of these uses.

General question 1

The electromagnetic spectrum is shown below.

electromagnetic spectrum						
radio and TV	microwaves	infrared	visible light	ultraviolet	X-rays	gamma rays

Different types of waves in the spectrum are used in medicine.

(a) What property do all electromagnetic waves have in common?

(b) Light can be produced by lasers. Describe the use of the laser in **one** application of medicine.

Go back to look at Chapter 1 if you need to revise the electromagnetic spectrum.

General question 1 – Answer

(a) *They all travel at the speed of light or at 3×10^8 m/s*

(b) *For cutting tissue/eye surgery/removing tattoos or birthmarks/sealing blood vessels.*

Don't be lazy and put 3×10^8 m/s alone or 'the same speed' alone as no marks will be awarded.

Use of X-rays

What you should know at General **and** Credit **level...**

X-rays are part of the electromagnetic spectrum with a shorter wavelength and increased energy compared to visible light (see p 112). Photographic film can be used to detect X-rays which is very useful in medicine where they are commonly used to identify broken bones. Bones absorb the X-rays but if a bone is broken, the X-rays pass through easily and show up on photographic film.

continued

A Computerised Tomography (CT) scanner is a sophisticated type of X-ray machine which takes images of horizontal slices through the body. The X-ray tube and detector rotate around the body. The information is analysed by a computer which produces a 3-D picture on a monitor. This method of imaging provides much more detailed information about the tissue, compared to conventional X-rays.

Learn this fact as it is often asked.

Uses of ultraviolet and infrared light in medicine

What you should know at **General** **level...**

Infrared radiation (IR) is part of the electromagnetic spectrum with longer wavelengths than visible light. It is produced by hot objects. Heat pictures (thermograms) of the human body can be used to detect areas which have abnormal heat patterns due to irregular blood flow. IR can also be used to treat muscle injury by heating the affected area and increasing the blood flow to aid healing.

Ultraviolet radiation (UV) is part of the electromagnetic spectrum with shorter wavelengths than visible light. UV can be used to treat skin conditions such as acne. However excessive exposure to UV radiation raises the risk of developing skin cancer.

Credit question 1

Gamma rays, ultraviolet and infrared are three members of a family of waves. Every member of this family travels at the speed of light.

(a) What name is given to this family of waves?

(b) Some uses of waves in this family are shown.

From the examples, give a use for:
(i) gamma rays (ii) ultraviolet (iii) infrared

(c) Which of the three waves in (b) has:
(i) the longest wavelength
(ii) the highest frequency?

Photographing bones inside a body

Tanning with a sun-ray lamp

Sterilising medical instruments

Communicating with a mobile phone

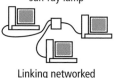
Linking networked computers through optical fibres

Treating injuries using a heat-lamp

Credit question 1 – Answer

(a) *Electromagnetic waves*

(b) (i) *gamma rays sterilising medical instruments*

　　(ii) *ultraviolet tanning with a sun lamp*

　　(iii) *infrared treating injuries using a heat lamp*

(c) (i) *longest wavelength infrared*

　　(ii) *highest frequency gamma*

Look out for

Learn the order of wavelengths in the electromagnetic spectrum.

Nuclear radiation – humans and medicine

The use of radioactivity in medicine

What you should know at General **level...**

Nuclear radiation can kill or change living cells. Because of this, doctors can use large doses to treat cancer tumours by killing the abnormal cells but they must be very careful to minimise the risk to healthy tissue. This is known as **radiotherapy**. Smaller doses of nuclear radiation can be used to detect abnormalities in the body – known as **radiodiagnosis**. This involves a patient being given a radioactive substance (called a **tracer**) and its path round the body is monitored with a gamma camera to detect abnormalities.

The gamma camera detects the gamma rays using a crystal which produces flashes of light (scintillations). These are converted to electrical pulses which are analysed by a computer and an image is displayed on a monitor.

Nuclear radiation can also be used to sterilise medical instruments by killing germs and bacteria.

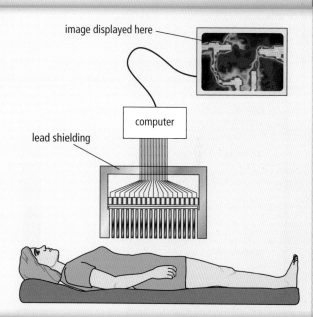

image displayed here

computer

lead shielding

The properties of radioactivity

What you should know at General **and** Credit **level...**

Nuclear radiation is energy which comes from the nucleus of an **atom**. An atom consists of a central nucleus containing protons and neutrons. Electrons orbit the nucleus.

Name	Charge (units)	Mass (units)
Proton	Positive +1	1
Neutron	Neutral 0	1
Electron	Negative −1	$\frac{1}{1836}$ (negligible)

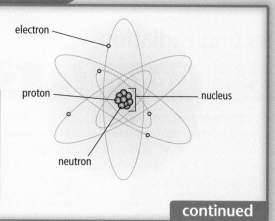

electron

proton

nucleus

neutron

continued

What you should know at General and Credit level – continued

There are 3 types of nuclear radiation – alpha, beta and gamma. You must be able to state their range and how they are absorbed.

Nuclear radiation	Symbol	What it is	Range in air	Absorbed by	Amount of Ionisation
alpha	α	A large positive nucleus	3 cm air	Sheet of paper	Very strong
beta	β	A fast moving electron	Metres of air	Thin piece of aluminium	Weak
gamma	γ	An electromagnetic wave	Air does not stop gamma	several cm lead or concrete	Very weak

Alpha rays produce much greater ionisation density than beta or gamma rays because of their large size.

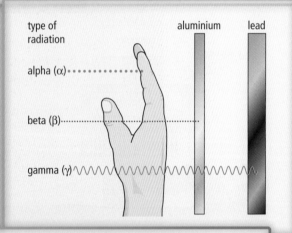

Look out for

Although **gamma radiation** travels the furthest and is the most penetrating radiation, alpha radiation does the most damage over a very short distance because of its strong ionising capability.

Also note that a **beta particle** is an **electron**. This is a particle usually found outside the nucleus. In beta decay protons are converted into electrons in the nucleus, producing beta particles.

Atoms usually have the same number of protons and electrons so an atom has no overall charge (it is neutral). However **ionisation** occurs when an atom gains an electron to give it an overall negative charge or when an atom loses an electron to give it an overall positive charge.

Look out for

Learn this definition carefully as it is asked frequently. You must include the word atom.

Detecting radiation

What you should know at General **and** Credit **level...**

Radiation can cause fogging of photographic film. The principle of fogging is used in **film badges** which are often used by workers who use radioactive material. When the film is developed, the amount of fogging and where it occurs can give an indication of what type and how much radiation the badge has been exposed to.

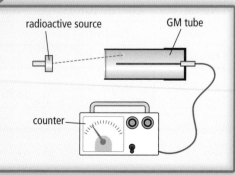

Another method for detecting nuclear radiation uses a Geiger-Muller (GM) tube: Radiation enters the tube and ionises the gas inside. This produces electrical pulses which are measured by a counter.

Credit question 1

As a safety precaution a technician wears a film badge when working with radioactive sources. The film badge contains photographic film. Light cannot enter the badge.

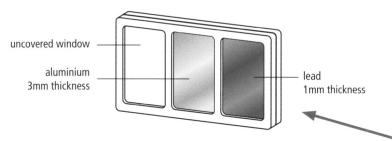

uncovered window

aluminium
3mm thickness

lead
1mm thickness

Describe how the film badge indicates the **type** and **amount** of radiation received.

Credit question 1 – Answer

Windows allow different radiations to pass through. Film becomes fogged/ blackened/ darkened.

Look out for

The uncovered window would allow all radiations to pass through. The aluminium would have stopped alpha radiation so beta and gamma radiation would be shown. The lead would block beta radiation, so gamma radiation would be the only one shown behind the lead window. The amount of fogging is a measure of the amount of radiation.

Radioactive decay and half-life

What you should know at **General** and **Credit** level...

When a radioactive source emits radiation, the source's radioactive atoms decay. The rate of decay is called the **activity** of the material and is measured in **becquerels (Bq)**:

1 becquerel (Bq) = 1 decayed atom per second

As atoms decay in the radioactive source, the activity of the source decreases with time. Some sources take thousands of years to decay whilst others remain active for only fractions of seconds. To compare the length of activity of substances the '**half-life**' is used. The half-life of a radioactive substance is the time taken for the activity to decrease by half.

Half-life can be measured using a GM tube attached to a counter placed in front of the radioactive source. The activity detected by the GM tube is measured over a period of time and a graph plotted as shown.

Look out for

The definition of a half-life is another definition which is often asked. It must be defined in terms of the activity and not **radioactivity**.

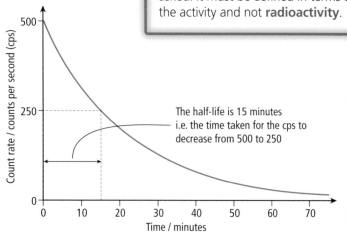

The half-life is 15 minutes i.e. the time taken for the cps to decrease from 500 to 250

General question 1

Radiotherapy treatment is provided by a machine which emits gamma radiation.

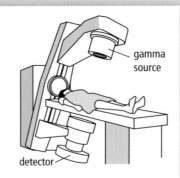

gamma source

detector

(a) Explain why gamma radiation is used rather than alpha or beta radiation.

(b) Explain why the gamma radiation is rotated around the patient.

General question 1 – Answer

(a) *gamma radiation penetrates body tissue (or alpha and beta would not reach the detector).*

(b) *maximum dose delivered or minimum dose to healthy tissue/focuses on cancer/healthy tissue is not damaged*

Credit question 1

A hospital technician is working with a radioactive source. The graph shows the activity of the source over a period of time.

(a) State what is meant by the term half-life.

(b) Use information from the graph to calculate the half-life of the radioactive source.

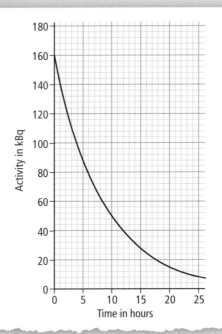

Credit question 1 – Answer

(a) *time for the activity (or number of nuclei)(of a radioactive source) to reduce to half the original number/activity/its value.*

(b) *Activity drops from 160 kBq to 80 kBq in 6 hours.*

   Half-life = 6 hours.

Look out for

Check this answer by working out how long it takes the activity to reduce from 100 kBq to 50 kBq.

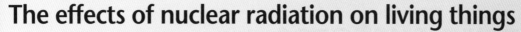

The effects of nuclear radiation on living things

What you should know at **General** level...

You have already learned that nuclear radiation can harm cells and tissue so safety precautions are essential when using radioactive sources.

Some of these are:

- Use forceps or a lifting tool to remove a source – never use bare hands.
- Keep the radiation source at a maximum distance from the body.
- Never bring a source close to your eyes.
- After any experiment with radioactivity, wash hands thoroughly.
- Radioactive sources must be stored in labelled boxes with appropriate shielding.

What you should know at **Credit** level...

For living materials, the biological effect of radiation depends on the absorbing tissue and the type of radiation. Some tissues and organs are more easily damaged by radiation (e.g. the eye) and alpha radiation does the most damage.

Why is this? Remember – alpha produces the greatest ionisation density (see page 54).

The **equivalent dose** measured in **sieverts** (Sv) takes account of the type and energy of radiation.

Credit question 1

The thyroid gland, located in the neck, is essential for maintaining good health.

(a) A radioactive source, which is a gamma radiation emitter, is used as a radioactive tracer for the diagnosis of thyroid gland disorders.
A small quantity of this tracer, with an activity of 20 MBq, is injected into a patient's body. After 52 hours, the activity of the tracer is measured at 1·25 MBq. Calculate the half-life of the tracer.

(b) Another radioactive source is used to treat cancer of the thyroid gland. This source emits only beta radiation. Why is this source unsuitable as a tracer?

(c) The equivalent dose is much higher for the beta emitter than for the gamma emitter. Why is this higher dose necessary?

(d) What are the units of equivalent dose?

Remember to include the unit 'hours'.

Credit question 1 – Answer

(a) $20 \rightarrow 10 \rightarrow 5 \rightarrow 2.5 \rightarrow 1.25$ MBq
 4 half lives = 52 hours ∴1 half life = 13 hours

(b) It is a beta emitter, absorbed within the body
 or gamma emitter required, to pass through body

(c) (larger dose required) to kill the (cancerous) cells

(d) Sievert (Sv)

Electronic systems
Input, process, output

What you should know at General **and** Credit **level...**

Electronic systems are all around you in everyday life. Computers, televisions, games consoles and mp3 players are all examples of electronic systems that people use every day.

Electronic systems can be broken down into three main parts:

INPUT → **PROCESS** → **OUTPUT**

A hi-fi is an example of an electronic system. The CD player is the input, taking the information from the CD and turning it into an electrical signal. The amplifier then processes this signal and the loudspeakers are the output, taking the processed electrical signal and turning it into a useful energy output as sound.

Electronic signals

What you should know at General **and** Credit **level...**

We can classify the signals in electronic systems as either **analogue** or **digital**. An **analogue** signal is one which varies continuously, and can have any value in a range, such as the one shown here:

A **digital** signal can either be **on** or **off**. We often refer to these states as being **high** or **low** and we can use the binary number **0** to represent low or off and **1** to represent high or on. An example of a digital signal is shown here:

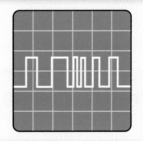

Output devices
Devices producing light, sound and movement

What you should know at General **and** Credit **level...**

You need to be able to suggest suitable output devices for applications, state the energy transformation and whether the device is analogue or digital. Some examples are given in the table below:

Output device	Useful energy out	Analogue or Digital
Lamp	Light	Can be both
LED	Light	Digital
Solenoid	Kinetic	Digital
Motor	Kinetic	Analogue
Loudspeaker	Sound	Analogue
Buzzer	Sound	Digital

! Look out for

The energy into an output device in a system is always **electrical energy**. Examiners will accept **electric energy** but answers like **electricity** or **electricity energy** will not be awarded any marks.

The LED

The **Light Emitting Diode (LED)** is one of the most widely used **output devices**. You will find them on most electrical equipment and these days high intensity LEDs are used in cars and large screen displays.

The circuit symbol for an LED is shown here: —▷|—

LEDs only conduct if they are connected the correct way round, as follows:

If you think of the LED symbol as looking like an arrowhead, then it should point towards the negative terminal of the supply. LEDs have a maximum working voltage and current, so need to be connected in series with a resistor in order to protect them from damage by limiting the current through the LED.

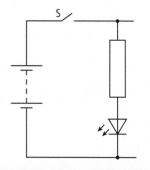

7-segment display

A 7-segment display is an output device, sometimes made up of 7 LEDs arranged in a figure 8 as shown. By lighting the appropriate segments, it is possible to produce any digit between 0 and 9.

A bank of 7-segment displays, such as the LCD screen on a calculator, can be used to display numbers with lots of digits.

General question 1

What output device could be used for a door opening and closing mechanism?

General question 1 – Answer

Solenoid

An electric motor or solenoid could be used for a door opening mechanism

Look out for

As with many other questions, the examiner is looking for one answer so don't be tempted to put more than one. Remember wrong answers cancel out correct ones!

General question 2

An electronic system is designed to count the number of vehicles that enter a car park. When a vehicle enters the car park, it cuts through a beam of light and a sensor circuit produces a digital pulse. The number of pulses produced by the sensor circuit is then counted and decoded before being displayed. The display consists of a number of illuminated sections.

(a) Name the device used to display the number of vehicles that enter the car park.

(b) The counter is reset to zero. Over a period of time, the sensor circuit then produces the following signal:

logic level 1
logic level 0

Shade in the sections that should be illuminated to show the number of vehicles that have entered the car park.

General question 2 – Answer

(a) 7-segment display

(b) There are four digital pulses so the seven

segment display would show the number 4.

Credit question 1

A torch contains five identical LEDs connected to a 3·0 V battery as shown:

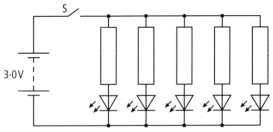

(a) State the purpose of the resistor connected in series with each LED.

(b) When lit, each LED operates at a voltage of 1·8 V and a current of 30 mA. Calculate the value of the resistor in series with each LED.

'The resistor connected in series limits current (through the LED)' or 'It prevents damage to the LED' are also acceptable answers

Answers such as 'to reduce voltage'; 'the voltage through'; to stop LED blowing'; 'to prevent LED from overheating' would not be accepted.

Credit question 1 – Answer

(a) It protects the LED from too much current.

(b) $V_R = 3·0 - 1·8 = 1·2$ V

$R = \dfrac{V}{I}$

$R = \dfrac{1·2}{0·03}$

$R = 40\,\Omega$

Questions on calculating the resistance required in series with an LED come up quite frequently. The most common mistake made is forgetting to calculate the voltage across the resistor first. This is done by subtracting the operating voltage of the LED from the supply voltage. In all questions like this it should be the first thing you do! The resistor and LED are connected in series, so the operating current for the LED is also the current in the resistor.

Input devices
Energy transformations in input devices

Input devices can be classified in two ways:

▶ devices which produce a voltage in response to a change such as light level, temperature or **force**, such as a microphone, thermocouple or solar cell.

▶ devices which change their resistance in response to a change such as light level or temperature such as a LDR or thermistor.

You need to be able to identify a suitable input device for a given application **and** identify the energy transformation for a microphone, thermocouple and solar cell.

Input device	Main energy transformation	Possible application
Microphone	Sound to Electrical	Karaoke machine
Thermocouple	Heat to Electrical	Monitoring high temperatures e.g. in a furnace
Solar cell	Light to Electrical	Power supply for a calculator

Some input devices can be used to change a voltage, and are often used as part of a **voltage divider**.

Switches are probably the simplest input device. When used in a voltage divider circuit they can make the voltage high or low depending upon where they are positioned.

Thermistors are devices which change resistance with temperature. Most thermistors are designed to have a decrease in resistance as the temperature rises, and the ones used in SQA questions are this type.

Look out for

Although some thermistors are designed to increase their resistance as temperature increases, all the ones used in SQA questions have been the type that shows a decrease in resistance. An easy way to remember how the resistance changes, is to remember LURD (light up resistance down) and TURD (temperature up resistance down).

Light Dependent Resistors (LDRs) are devices which are designed to a have a decrease in resistance as the light level increases.

Capacitors are also used in voltage divider circuits. A **capacitor** is a device which stores charge. As a capacitor charges the voltage across it increases, and when it is discharged the voltage across it decreases. The length of time a capacitor takes to charge depends upon the capacitance of the capacitor and the resistance of the resistor in series with it. The greater the value of the capacitance or resistance the longer the capacitor takes to charge. Capacitors are often used in timing circuits.

Credit question 1

A digital camera is used to take pictures. When switched on, the flash requires some time before it is ready to use. When ready, a green LED lights up.

digital camera

The part of the circuit used to control the LED is shown below. The voltage at point X is initially 0 V.

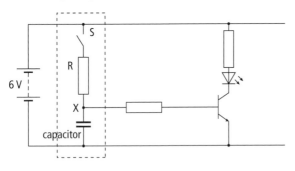

Note for these parts of the question we are only interested in the part of the circuit that has been highlighted for you.

(a) Describe what happens to the voltage at point X when switch S is closed.

(b The camera manufacturer wants to change the time taken for the flash to be ready to operate. State **two** changes which could be made to the above circuit so that the time taken for the green LED to come on is **reduced**.

Credit question 1 – Answer

(a) *The voltage gradually increases.*

(b) *Reduce the resistance (or reduce the value of the resistor) and*
reduce the capacitance.

Look out for !

This is another example of a question where you need to be precise in the language you use. Writing answers such as 'use a smaller resistor or capacitor' or 'lower the resistor or capacitor' will score 0 marks, as they are too imprecise. The physical size of a resistor is not related to its resistance, so a 'smaller resistor' may not have a **smaller resistance**; similarly 'lower the resistor' could mean move its position to lower in the circuit rather than **lower resistance**.

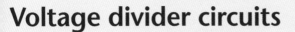

Voltage divider circuits

What you should know at **General** **and** **Credit** **level...**

Voltage divider circuits are often used to divide up a voltage, for example from a power supply. This can be done using resistors as shown:

There are two ways we can find voltage V_1.

Method 1 We can use the relationship:

$$V_1 = \left(\frac{R_1}{R_1 + R_2}\right)V_s$$

Method 2 We can also use Ohm's Law and the fact that the current through each resistor is the same, as they are connected in series:

$$I = \frac{V_1}{R_1} = \frac{V_2}{R_2} \quad \text{so} \quad \frac{V_1}{R_1} = \frac{V_2}{R_2}$$

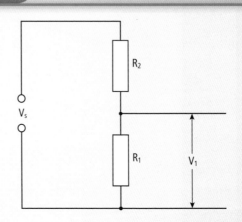

Voltage dividers are often constructed with one of the resistors being a device which changes its resistance with a change in the environment, such as an LDR or thermistor.

For the circuit on the right, if the light level **decreased** the resistance of the LDR would **increase** and so the voltage across it, V_1, would also increase, as a voltage divider shares the voltage across the resistors in ratio with their resistances.

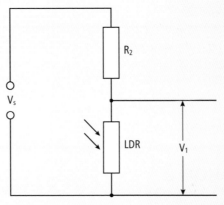

Credit question 1

The camera flash on page 62 is designed to operate under dim lighting conditions. Another part of the circuit for the flash is shown below. The flash only operates when a minimum voltage of $0\cdot7$ V occurs across the LDR.

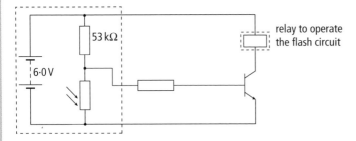

(a) Calculate the voltage across the 53 kΩ resistor when the voltage across the LDR is $0\cdot7$ V.

(b) Calculate the **minimum** resistance of the LDR that allows the flash to operate in dim conditions.

Look out for

This is actually a straightforward question which makes use of the fact that the resistor and LDR are connected in series and so the voltage divides across them. You need to focus only on the supply and voltage divider parts of the circuit shown in the dashed box.

Credit question 1 – Answer

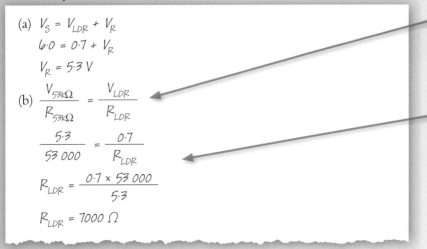

(a) $V_S = V_{LDR} + V_R$

$6 \cdot 0 = 0 \cdot 7 + V_R$

$V_R = 5 \cdot 3 \ V$

(b) $\dfrac{V_{53k\Omega}}{R_{53k\Omega}} = \dfrac{V_{LDR}}{R_{LDR}}$

$\dfrac{5 \cdot 3}{53\ 000} = \dfrac{0 \cdot 7}{R_{LDR}}$

$R_{LDR} = \dfrac{0 \cdot 7 \times 53\ 000}{5 \cdot 3}$

$R_{LDR} = 7000 \ \Omega$

There are a number of ways you can do this question, but the easiest way is shown here.

In this question, you need to use your answer from the first part to get the voltage across the resistor and the stem of the question tells you that the minimum voltage across the LDR has to be 0·7 V and R is 53 kΩ. Once you've put the numbers in the correct place you need to rearrange to make R_{LDR} the subject of the equation.

Look out for !

Remember, unless you are confident at rearranging equations, put the numbers into the equation first and then rearrange. If the examiners can see that you've substituted the values into the correct place in the equation but have then made a mistake in working out your answer you would only lose ½ mark for an arithmetic mistake. If you rearrange the equation incorrectly then the maximum you could obtain is ½ mark for selecting the correct formula.

Credit question 2

A high intensity LED is used as a garden light. The light turns on automatically when it becomes dark. The light also contains a solar cell which charges a rechargeable battery during daylight hours. Part of the circuit is shown below:

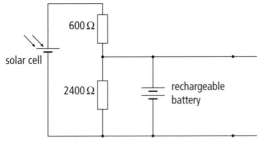

(a) State the energy transformation in a solar cell.

(b) At a particular light level, the voltage generated by the solar cell is 1·5 V. Calculate the voltage across the rechargeable battery at this light level.

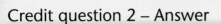

Credit question 2 – Answer

(a) <u>Light energy to electrical energy</u>

(b) $V_1 = \left(\dfrac{R_1}{R_1 + R_2}\right)V_S$

$V_1 = \left(\dfrac{2400}{2400 + 600}\right) 1\cdot5$

$V_1 = 1\cdot2\,V$

Again, there are a number of ways you can do this question but the easiest way is as shown:

Be careful to match the voltage you're finding, V_1, with the correct resistor. In this case, it is the 2400 Ω resistor.

Look out for

There are often questions which test your understanding of a voltage divider circuit, many of these involve calculations but some will involve explanations about how a transistor switching circuit operates. There are a number of approaches you can take to the calculations, if you can't remember or select the correct formula for the potential divider then you can use Ohm's Law calculations and treat the voltage divider as a series circuit.

Digital processes
Transistor switching circuits

What you should know at **General** and **Credit** **level...**

Transistors are important electrical components that feature in many electrical devices you have, including your computer, mp3 player and television. Sometimes they are discrete components like the ones shown in the photo, but often they are contained in a microchip.

A **transistor** is a **semiconductor** device which can be used as a switch. There are lots of different types of transistor, but the only one you need to know about at Standard Grade is called an NPN transistor.

The circuit symbol for an NPN transistor is shown here:

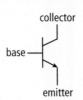

collector

base

emitter

The names of the three connections – base, collector and emitter – are not part of the symbol but are useful to know.

When a transistor is used as a switch in a circuit, the transistor is off (non-conducting) until the voltage between the base and emitter reaches a certain level (typically 0·7 V for an NPN transistor) at which point the transistor switches on (starts conducting).

Credit question 1

An electronic circuit is shown below. Component R is a thermistor.

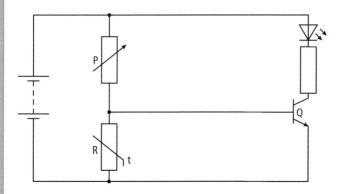

(a) Name component P

(b) Name component Q.

(c) In this circuit, what is the function of component Q?

(d) Explain how the circuit operates.

Credit question 1 – Answer

(a) *Variable resistor*

(b) *(NPN) transistor*

(c) *The transistor acts as a switch.*

(d) *As the temperature decreases, the resistance of the thermistor increases. The voltage across the thermistor increases and when it reaches 0·7 V the transistor switches on, which switches on the LED.*

You don't need to include NPN in your answer; at Standard Grade 'transistor' on its own is sufficient.

You don't have to state '0·7 V'; you could also say 'a certain voltage' or 'a set voltage'.

Look out for

Occasionally questions will ask you to suggest a use for transistor switching circuits. A circuit like this could be used in an incubator, for example, to warn if the temperature fell below a certain value.

Look out for

In a transistor switching circuit like this one, swapping the positions of the thermistor and variable resistor would reverse the way the circuit works, i.e. the LED would come on if the temperature went above a certain value.

Look out for

Questions asking you to explain how transistor switching circuits work come up reasonably frequently. Be methodical in your description; start at the left hand side of the circuit with the voltage divider and explain how it works, then work your way to the right of the circuit through the transistor to the output.

Logic gates

What you should know at General **and** Credit **level...**

Logic gates are digital devices found in the microchips used in many electrical devices including computers and calculators.

There are many types of logic gates but for Standard Grade Physics you only need to know three.

▶ **NOT** gate (also called an **inverter**)

▶ **OR** gate

▶ **AND** gate.

The NOT gate has one input but other logic gates can have many inputs. You will only need to work with NOT gates and with AND gates and OR gates that have two inputs.

The circuit symbols for the logic gates are given below:

NOT gate or Inverter

The Input and Output labels are not required as part of the symbol.
2-input OR gate

2-input AND gate

Look out for

Questions about logic gates come up frequently both in the General and Credit papers. Combinations of logic gates can be used in circuits for many applications. Some of the examples you will see in Past Paper questions include burglar alarms, drink vending machines and security systems in banks.

Truth tables can be used to show the outputs for each possible combination of input:

▶ **1** is used to represent **high** or **on**

▶ **0** is used to represent **low** or **off**.

NOT gate

Input	Output
0	1
1	0

An easy way to remember how a NOT gate works is that the output is **NOT** the same as the input.

OR gate

Input		Output
A	B	
0	0	0
0	1	1
1	0	1
1	1	1

For an OR gate, the output is high if one input **OR** the other input **OR** both inputs are high.

AND gate

Input		Output
A	B	
0	0	0
0	1	0
1	0	0
1	1	1

For an AND gate, the output is only high if input A **AND** input B are high.

General question 1

An electronic control system is used to control a lift. When a floor has been selected, two checks are made:

- there are no obstructions to the doors;
- the lift is not overloaded.

Part of the circuit is shown below.

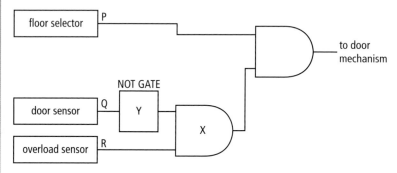

The logic states are as shown for the floor selector, the sensors and the door mechanism.

		logic level
floor selector	not pressed	0
	pressed	1
door sensor	no obstruction	0
	obstruction	1
overload sensor	overloaded	0
	not overloaded	1
door mechanism	doors open	0
	doors closed	1

(a) Name logic gate X.

(b) Gate Y is a NOT gate. State the logic levels needed at P, Q and R to close the lift doors.

General question 1 – Answer

(a) *AND gate*

(b) *P = 1*

    Q = 0

    R = 1

This question can be done in two ways; The first is to think about the conditions required i.e. the floor selector needs to be pressed (1), there should be no door obstruction (0) and the lift should not be overloaded (1). Alternatively, you can work backwards from the output of the combination of gates. The output of the last AND gate is a 1, so both inputs have to be a 1, which means P is 1. The output of the first AND gate also has to be 1, so both inputs are a 1, so R is 1. The output of the NOT gate is 1 so the input Q must be 0.

Credit question 1

The exit of an underground car park has an automatic barrier. The barrier rises when a car interrupts a light beam across the exit and money has been put into the pay machine. The barrier can also be operated using a manual switch.

Part of the control circuit for the automatic barrier is shown.

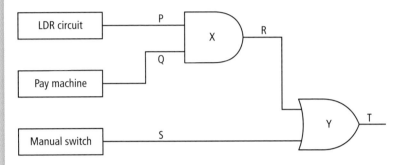

When a car interrupts the light beam, the logic level at P changes from logic 0 to logic 1.

When money is put into the pay machine, the logic level at Q changes from logic 0 to logic 1.

When the manual switch is operated, the logic level at S changes from logic 0 to logic 1.

(a) Name logic gate X.

(b) Name logic gate Y.

(c) Complete the truth table for the control circuit shown to show the values of logic levels at R and T.

P	Q	R	S	T
0	0		0	
0	1		0	
1	0		0	
1	1		0	
0	0		1	
0	1		1	
1	0		1	
1	1		1	

Credit question 1 – Answer

(a) AND gate

(b) OR gate

(c)

P	Q	R	S	T
0	0	0	0	0
0	1	0	0	0
1	0	0	0	0
1	1	1	0	1
0	0	0	1	1
0	1	0	1	1
1	0	0	1	1
1	1	1	1	1

R is the output from an AND gate so requires both P **AND** Q to be high (1). T is the output from an OR gate so will be high (1) if either R **OR** S **OR** both are high (1).

Clock pulse generator

What you should know at **General** and **Credit** level...

Many electronic devices, such as computers, require internal clocks so that they function correctly. They use a digital circuit called a **clock pulse generator** which consists of a capacitor, resistor (or variable resistor) and a NOT gate (inverter).

The frequency of the clock pulses can be altered by changing the value of the capacitance or resistance in the circuit. Increasing the resistance or the capacitance will decrease the frequency of the pulses.

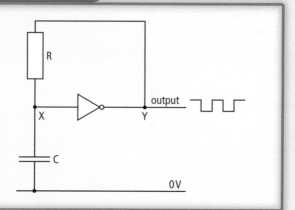

Credit question 1

(a) Capacitor C is initially discharged. Explain the operation of the pulse generator circuit, by referring to points X and Y in the circuit.

(b) The pulse generator produces an output with a high frequency. State **one** change that could be made to give an output of lower frequency.

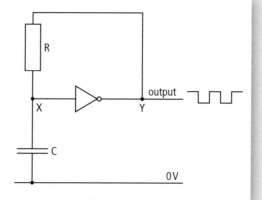

Credit question 1 – Answer

(a) *Initially the capacitor is discharged so the voltage at point X is 0 V and the input to the NOT gate is low. The output Y of the NOT gate is high so the capacitor charges through the resistor. As the capacitor charges the voltage across it rises and when the voltage at point X becomes high, the input to the NOT gate is high so the output Y becomes low. The capacitor now discharges through the resistor and when the voltage at point X becomes low the output of the gate goes high. The cycle then repeats.*

(b) *The frequency of the clock pulses could be lowered by increasing the capacitance (or resistance).*

Look out for

Explaining how a clock pulse generator works is a Credit level outcome. As with previous questions, try and be methodical in your explanation.

This is another question where you need to be precise about the language you use. Remember you need to avoid writing answers like 'use a bigger capacitor or resistor'; you should make sure you use the proper terms 'capacitance' or 'resistance'.

Counters

What you should know at General **and** Credit **level...**

The pulses from the generator can be counted using a **binary counter**. The output of this counter is in binary so a decoder is then required to convert the binary number into decimal, which can then be displayed on a 7-segment display.

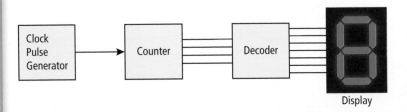

Display

Analogue processes

Amplifiers

What you should know at General **and** Credit **level...**

Amplifiers are found in many electronic systems, such as hifis, mp3 docking stations, televisions, radios, intercoms – in fact any system where a small signal needs to be 'made bigger' or amplified.

The **voltage gain** of an amplifier can be calculated using the relationship:

$$V_{gain} = \frac{V_O}{V_i}$$

Where V_{gain} is the voltage gain, V_O is the output voltage in volts (V) and V_i is the input voltage in volts.

The power output of an amplifier can be calculated using the relationship:

$$P = \frac{V^2}{R}$$

Where P is the power in watts (W), V is the voltage in volts (V) and R is the resistance in ohms (Ω).

The **power gain** of an amplifier can be calculated using the relationship:

$$P_{gain} = \frac{P_O}{P_i}$$

Where P_{gain} is the power gain, P_o is the output power in watts (W) and P_i is the input power in watts (W).

General question 1

A radio controlled model fire engine receives signals from a control unit. One of the control functions operates a siren on the fire engine. The fire engine contains an electronic system to control the siren. The signals at various parts of the system are displayed on oscilloscope screens.

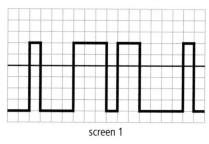
screen 1

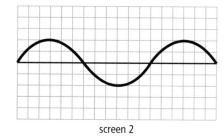
screen 2

(a) Which screen shows a digital signal?

(b) The signal shown on screen 2 is now amplified. The oscilloscope settings are unchanged. Draw the amplified signal.

General question 1 – Answer

(a) _screen 1_

(b)
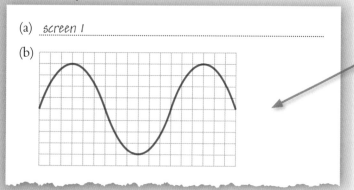

The amplified signal should have greater amplitude than the original signal. It is difficult when drawing freehand to be really accurate so examiners allow you some tolerance in your drawing but you still need to be as neat and careful as possible, so use the grid lines as a guide for your drawing. In this example, a small tolerance was allowed but your answer had to show two crests and one trough and the amplitude had to be bigger than two boxes.

Credit question 1

An electronic tuner for a guitar contains a microphone and an amplifier. The output voltage from the amplifier is 9 V.

(a) The voltage gain of the amplifier is 150. Calculate the input voltage to the amplifier.

(b The tuner is used to measure the frequency of six guitar strings. The number and frequency of each string is given in the table. String 3 is plucked. What is the frequency of the output signal from the amplifier?

Number of string	Frequency (Hz)
1	330·0
2	247·0
3	196·0
4	147·0
5	110·0
6	82·5

Credit question 1 – Answer

(a) $V_{gain} = \dfrac{V_o}{V_i}$

$150 = \dfrac{9}{V_i}$

$V_i = \dfrac{9}{150}$

$V_i = 0.06\ V$

(b) *The frequency of the output signal is 196.0 Hz.*

Look out for

An amplifier **does not change the frequency** of the signal only the amplitude.

A very common mistake made by students is to multiply the frequency by the voltage gain. However, if you think carefully about this, if an amplifier increased the frequency of the signal all your favourite singers would sound as though they were singing on helium!

Credit question 2

An electronic device produces a changing light pattern when it detects music, but only when it is in the dark. The device detects music from a CD player. The CD player contains an amplifier that produces an output voltage of 5·6 V when connected to a loudspeaker of resistance 3·2 Ω.

(a) Calculate the output power of the amplifier.

(b) The input power to the amplifier is 4·9 mW. Calculate the power gain of the amplifier.

(c) One particular signal from the CD to the amplifier has a frequency of 170 Hz. What is the frequency of the output signal of the amplifier?

Credit question 2 – Answer

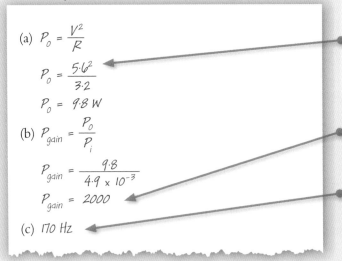

(a) $P_o = \dfrac{V^2}{R}$

$P_o = \dfrac{5.6^2}{3.2}$

$P_o = 9.8\ W$

(b) $P_{gain} = \dfrac{P_o}{P_i}$

$P_{gain} = \dfrac{9.8}{4.9 \times 10^{-3}}$

$P_{gain} = 2000$

(c) 170 Hz

Don't forget to include the square sign when substituting in your values in the second line. Examiners are very strict about this; even if you actually remember to square it when working out the answer on your calculator, if you don't include it you will lose most of the marks.

Remember there are no units for power gain.

As with the last example, the amplifier does not change the frequency of the signal.

Look out for

The last example illustrates something that often comes up in Physics exams. Part (b) requires you to use your answer from part (a) in the calculation. If you don't know how to do part (a), then write an answer into the part (a) box and use this number to do part (b). You will gain marks for part (b) if you use your wrong answer from part (a) correctly in part (b).

On the move

Average and instantaneous speed

What you should know at **General** **level...**

The **average speed** $\bar{v}$ of a moving object (e.g. a car) is calculated using two measurements and substituting these two measurements into the formula for average speed.

The measurements are:

1 the distance d between two lampposts (or any two convenient markers)

2 the time t taken to travel this distance (measured with a stopwatch)

The average speed is calculated using:

$$\bar{v} = \frac{d}{t}$$

The line (or bar) above the v is a shorthand way of saying **average speed**. The data booklet lists this relationship as:

$$d = \bar{v} t$$

Rearranging this formula to calculate t gives:

$$t = \frac{d}{\bar{v}}$$

There is no penalty if you don't include the bar above the v.

The **instantaneous speed** of a moving object is the average speed over a small time interval (usually much less than 1 second).

A light gate is often used to measure this small time interval when the card cuts the light beam.

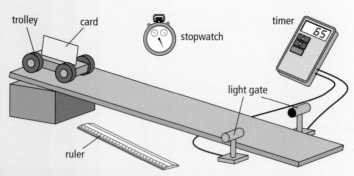

The instantaneous speed of the trolley at the light gate is:

$$\bar{v} = \frac{card\ width}{interrupted\ time}$$

The average speed of the trolley down the slope is:

$$\bar{v} = \frac{total\ length\ of\ slope}{time\ taken\ by\ trolley\ to\ travel\ down\ slope}$$

Look out for

When both the card width **and** the slope length are given great care must be taken to use the correct substitution for d in the formula

$$\bar{v} = \frac{d}{t}.$$

A common mistake is using the slope length when calculating instantaneous speed.

Average and instantaneous speed

You should be able to explain why the instantaneous speed of a car is often different from the average speed of the car. This is because the average speed is made up of many different speeds during the car's journey. The instantaneous speed of the car as it passes a sign post is the speed **over a short time interval** when the speed doesn't change much.

General question 1

An indoor kart track hosts a racing competition.

Describe how to find the average speed of a kart for one complete lap of the track. You must state the measurements that are made and how they are used.

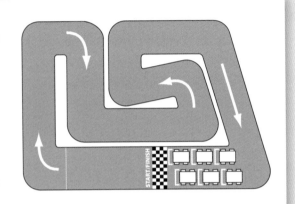

(3)

General question 1 – Answer

Measure the length of one lap of the track to determine the distance.

Measure the time taken by the kart to complete one lap of the track.

Divide the distance by the time to calculate the average speed of the kart.

Double check 3 mark questions and make sure you have 3 parts in your answer – many students will give the first two statements but forget the third statement.

General question 2

A rowing crew takes part in a race. The time for their boat at each stage of the race is shown in the table.

Calculate the average speed of the boat during the first 500 metres of the race. (2)

		Time from start	
		minutes	seconds
Start:	0 metres	00	00
	500 metres	01	40
	1000 metres	03	50
	1500 metres	05	50
Finish	2000 metres	07	45

General question 2 – Answer

$$\bar{v} = \frac{d}{t}$$

$$= \frac{500}{100}$$

$$= 5\,m/s$$

*The time is given as 1 minute 40 seconds in the table. This has to be changed into seconds as the answer has to be in **metres per second**.*

Acceleration

What you should know at **General** and **Credit** level...

Acceleration occurs when an object's speed changes. Constant acceleration *a* can be calculated using:

$$a = \frac{v - u}{t}$$

where   *u* = initial speed
   v = final speed
   t = the time taken

Acceleration is the change in speed *per second* and has a unit of *metres per second per second* or m/s². An acceleration of 3 m/s² means the speed is increasing by 3 m/s every second.

An acceleration of –4 m/s² means the speed decreases by 4 m/s every second.

Credit question 1

A bobsleigh team competes in a race.

(a) Starting from rest, the bobsleigh reaches a speed of 11 m/s after a time of 3·2 s. Calculate the acceleration of the bobsleigh.   (2)

(b) The bobsleigh completes the 1200 m race in a time of 42·0 s. Calculate the average speed of the bobsleigh.   (2)

(c) Describe how the instantaneous speed of the bobsleigh could be measured as it crosses the finishing line.   (2)

Credit question 1 – Answer

(a) $a = \dfrac{v - u}{t}$

$= \dfrac{11 - 0}{3 \cdot 2}$

$= 3 \cdot 4 \text{ m/s}^2$

> The unit of acceleration is not the same as the unit of speed. Many exam calculations for acceleration have the correct numerical answer but have the wrong unit because the square part is missing from m/s^2. This will lose ½ mark out of the 2 marks allocated

(b) $v = \dfrac{d}{t}$

$= \dfrac{1200}{42}$

$= 28 \cdot 6 \text{ m/s}$

(c) <u>set up a light gate at the finish line</u>

<u>measure the length of the bobsleigh (which cuts the light beam)</u>

<u>note the interrupted time (from the timer connected to the light gate)</u>

$\text{instantaneous speed} = \dfrac{\text{length of bobsleigh}}{\text{interrupted time}}$

> Remember to include the formula for instantaneous speed in your answer.

Speed-time graphs

What you should know at **General** level...

The speed-time graph of a moving object gives information about the speed of the object during a particular time interval.

A car starts from rest and accelerates up to a steady speed of 12 m/s in a time of 4 seconds. The car travels at 12 m/s for a further 7 seconds before decelerating to rest in a time of 2 seconds. The speed time graph of the whole journey will look like this:

Look out for

> Be careful not to substitute 13 seconds for the time of the initial acceleration – a common mistake.

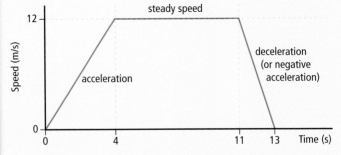

The initial acceleration (between zero and 4 seconds) can be calculated as follows:

$a = \dfrac{v - u}{t} = \dfrac{12 - 0}{4} = 3 \text{ m/s}^2$

What you should know at **Credit** level...

At Credit level, the area under a speed time graph is used to calculate the distance travelled.

A car's journey is represented by the graph.

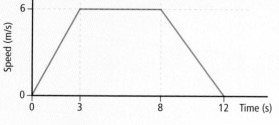

The distance travelled by the car during the first 3 seconds of the journey is the area of the first triangle.

$Area = \frac{1}{2}$ base × height = 0·5 × 3 × 6 = 9 m

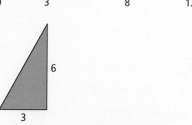

The distance travelled during the whole journey lasting 12 seconds would be the total area under the graph. The area is broken down into two triangles and a rectangle.

$Area$ = 0·5 × 3 × 6 + 5 × 6 + 0·5 × 4 × 6
      = 9 + 30 + 12 = 51 m

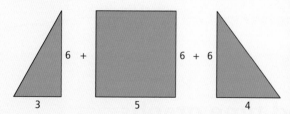

Credit question 1

A jet-engined car of mass 10 000 kg was used to set a land-speed record. The graph shows the speed of the car during the first part of one test run.

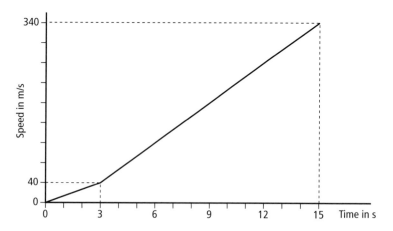

(a) Calculate how far the car travelled in the 15 s shown. (3)

(b) Calculate the maximum acceleration of the car during this part of the test run. (2)

An exam question asking for distance travelled may be disguised slightly. The speed–time graph of an aircraft during takeoff might be given and you are asked for the length of the runway. This will be equal to the area under the speed–time graph.

Credit question 1 – Answer

(a) <u>Distance travelled = area under the speed time graph</u>

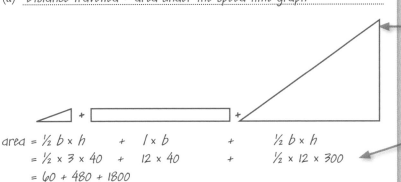

area = ½ b × h   +   l × b   +   ½ b × h
     = ½ × 3 × 40 + 12 × 40   +   ½ × 12 × 300
     = 60 + 480 + 1800
     = 2340 m

(b) <u>The acceleration from 3 s to 15 s is greater than the acceleration</u>

<u>from 0 to 3 s.</u>

$$a = \frac{v - u}{t}$$

$$= \frac{340 - 40}{12}$$

$$= \frac{300}{12}$$

$$= 25 \ m/s^2$$

<u>The maximum acceleration will have the steepest slope on the</u>

<u>speed-time graph as the greatest increase in speed per second</u>

<u>happens here.</u>

> The area under the graph must be broken down into triangles and rectangles whose area can be calculated.
> The area in this question consists of 2 triangles and a rectangle.

> The height of the bigger triangle is 300 not 340. The triangle doesn't start until 40 up from the time axis.
> Similarly the length of the rectangle is 12 not 15.

Look out for

The Credit exam always has a question where the area under a speed-time graph has to be calculated. The question however will simply ask for a distance travelled (not an area).

Forces at work

Forces, weight and mass

What you should know at General **level...**

A force applied to an object can change the object's **shape, speed and direction** of travel e.g. pulling a length of plasticine or pushing a sledge.

The size of a force is measured in **newtons** (N) and can be measured using a **newton balance.**

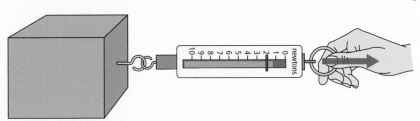

continued

What you should know at General level – continued

Weight

The weight of an object is the force of gravity on the object and can be calculated using the relationship:

$W = m \times g$,   where W = the weight of the object in newtons (N)
m = the mass of the object in kilograms (kg)
g = the gravitational field strength (10 N/kg on Earth)

Example: Calculate the weight of a fish whose mass is 2·3 kg.

$W = m \times g = 2\cdot3 \times 10 = 23$ newtons

What you should know at Credit **level...**

Mass and weight are **different** quantities in physics. Mass is the amount of matter in an object measured in kg. Weight is the gravitational force that a **planet** exerts on an object measured in N.

Consider two astronauts of masses 60 kg and 80 kg.

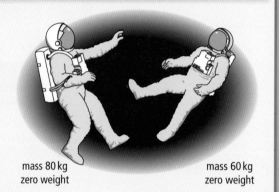

mass 80 kg
zero weight

mass 60 kg
zero weight

In deep space the weights of these two astronauts will be zero. But their masses will still be 80 kg and 60 kg. There will be more kilograms of mass in the bigger astronaut even although both are weightless.

Gravitational field strength

Gravitational field strength is defined as the **weight per unit mass** (or the force of gravity acting on each kilogram). On Earth the gravitational field strength is 10 newtons per kilogram.

General question 1

A car ferry has a mass of 1 600 000 kilograms.
What is the weight of the ferry?   (2)

General question 1 – Answer

$W = mg$

$= 1\ 600\ 000 \times 10$

$= 16\ 000\ 000$ N

> **Look out for**
>
> In everyday (non-physics) life, mass and weight are often taken to be the same thing. This simplification succeeds because most of us never leave the region on or near the Earth's surface so the weights and masses of objects stay constant.

> **Look out for**
>
> Don't forget the **per kilogram** part to your explanation or you will not get the mark if asked to explain what is meant by **gravitational field** strength. This is a very popular credit question.

Friction

The force of friction **opposes** the motion of an object. **Air resistance** or **drag** is a form of friction and acts in the direction **opposite** to a vehicle's motion.

Friction can be decreased by **streamlining** the shape of a vehicle's bodywork so the air passes over more smoothly.

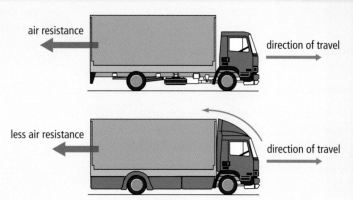

This will improve the fuel consumption of the lorry.

The friction on a vehicle is deliberately increased when the lorry brakes. Brake pads grip part of the wheel during braking creating friction.

Balanced forces – Newton's First Law

When the forces on a moving object are **equal and opposite** we say the forces are **balanced**.

An example of balanced forces is

Newton's First Law states:

> If the forces on a moving object are **balanced** then the speed of the object stays **constant**.

The reverse of this is also true:

> If the speed of a moving object is **constant** then the forces acting on the object must be **balanced**.

For example, a 65 kg parachutist descends vertically at a steady speed of 6 m/s. The weight of the parachutist downwards (650 N) must be **the same** as the air resistance (frictional force) upwards.

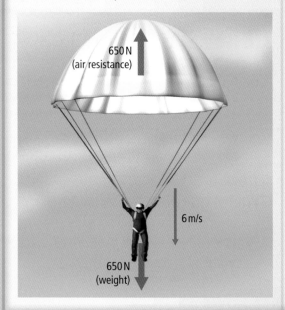

Unbalanced forces – Newton's Second Law

Consider this situation where the forces on an object are not balanced.

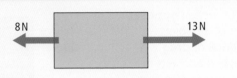

8 N   13 N

The resultant or **unbalanced force** on the object is 5 N to the right.

This unbalanced force causes the object to accelerate to the right and Newton's Second Law states:

$F = ma$ where F = the unbalanced force on the object (in N)
            m = the mass of the object (in kg)
            a = the acceleration of the mass (in m/s²)

Example: Calculate the acceleration of a 7·5 kg object when an unbalanced force of 5·2 N is applied.

$$a = \frac{F}{m} = \frac{5 \cdot 2}{7 \cdot 5} = 0 \cdot 69 \text{ m/s}^2$$

Numerical calculations involving $F = ma$ (Newton's Second Law)

You are expected to be able to attempt Newton's Second Law calculations where more than one force is applied.

7 N   3 kg   18 N

Example: Calculate the acceleration of the 3 kg mass in the diagram.

$$a = \frac{F}{m} = \frac{18 - 7}{3} = \frac{11}{3} = 3 \cdot 67 \text{ m/s}^2$$

General question 1

(a) A car ferry with a weight of 16 000 000 newtons floats at rest in the harbour. The water applies an upward force to the ferry. Is this upward force greater than, less than, or equal to the weight of the ferry? (1)

(b) When the engines are started, the propellers of the ferry apply a force of 100 000 newtons to the ferry. Calculate the initial acceleration of the ferry. (2)

General question 1 – Answer

The forces must be balanced as the ferry floats.

(a) *upward force is **equal** to the weight*

(b) $a = \dfrac{F}{m} = \dfrac{100\ 000}{16\ 000\ 000} = 0 \cdot 00625 \text{ m/s}^2$

General question 2

A helicopter of mass 12 000 kilograms travels to the scene of an accident to rescue an injured climber. The forces acting on the helicopter are as shown in the diagram:

(a) The helicopter travels at constant speed. State how the forward force compares with the air friction force during the journey. (1)

(b) Helicopter designers try to reduce the air friction as much as possible. How can they do this? (1)

(c) Calculate the weight of the helicopter. (2)

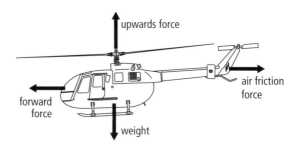

General question 2 – Answer

(a) Forward force is the same as the friction force.

(b) Streamline the shape of the helicopter.

(c) W = mg = 12 000 × 10 = 120 000 N

This is a **state** question so no explanation is required. (The explanation is that the forces are balanced as the helicopter has a steady speed.)

Credit question 1

A model motor boat of mass 4 kg is initially at rest on a pond. The boat's motor, which provides a constant force of 5 N, is switched on. As the boat accelerates, the force of friction on it increases. A graph of the force of friction acting on the boat against time is shown:

(a) State the force of friction acting on the boat 2 s after the boat starts after the motor is switched on. (1)

(b) Calculate the acceleration of the boat at this time. (3)

(c) Describe and explain the movement of the boat after 7 s. (2)

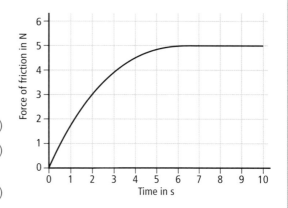

Credit question 1 – Answer

(a) Force of friction = 3 N

(b) unbalanced force = motor force – friction force

= 5 – 3

= 2 N

$a = \frac{F}{m} = \frac{2}{4} = 0.5 \, m/s^2$

(c) The boat has a steady speed because the forces are balanced (both motor force and friction force are 5N).

Note this is a 3 mark calculation question. This will involve more than one calculation:

Describe and explain questions have 2 marks. Forgetting to complete the **explain** part is a common omission.

Credit question 2

An aeroplane on an aircraft carrier must reach a maximum speed of 70 m/s to safely take off. The mass of the plane is 28 000 kg.

The aeroplane accelerates from rest to its minimum take off speed in 2 s.

(a) Calculate the acceleration of the aeroplane. (2)

(b) Calculate the force required to create this acceleration. (2)

(c) The aeroplane's engines provide a total thrust of 240 kN. An additional force is supplied by a catapult to produce the acceleration required. Calculate the force produced by the catapult. (1)

Credit question 2 – Answer

(a) $a = \dfrac{v - u}{t} = \dfrac{70 - 0}{2} = 35 \text{ m/s}^2$

(b) $F = ma$

$ \doteq 28\ 000 \times 35$

$= 980\ 000 \text{ N}$

(c) extra force required by catapult $= 980\ 000 - 240\ 000$

$= 740\ 000 \text{ N}$

> Remember you must use the correct unit for acceleration: m/s² (not m/s)

Car seat belts

What you should know at General and Credit level...

Car seat belts are an important safety feature in motor travel. The physics behind their operation uses Newton's First and Second law.

(a) Without seat belt on. When the car and driver are travelling at a steady speed the horizontal forces acting on the driver are balanced. (Newton's First Law).

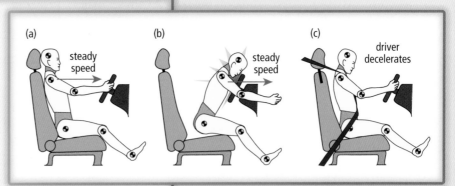

(a) steady speed
(b) steady speed
(c) driver decelerates

(b) Without seat belt on. When the car brakes, the brake force decelerates the car and seat. But the driver will continue to **move in a straight line at the same steady speed** as the brake force does not act on the driver. (Newton's First Law)

(c) With seat belt on. The seat belt provides an **unbalanced force** to **decelerate** the driver. (Newton's Second Law)

Look out for

A common mistake is to suggest that the driver moves forward due to some (unbalanced) force acting on the driver. Sounds plausible but this is wrong physics.

Movement means energy

Energy transformations

What you should know at General **level...**

Energy can be **changed** from one form to another. Consider a cyclist accelerating on a level road then freewheeling up then down a small hill. The various energy changes are described.

AB chemical energy to kinetic energy as cyclist accelerates from rest
BC kinetic energy to gravitational potential energy
CD gravitational potential energy to kinetic energy
DE kinetic energy to heat energy as cyclist brakes and decelerates

Work done

Work done is another name for the amount of **energy required** when a force F is applied over a distance d. The relationship is:

$E_w = F \times d$ where E_w is the work done (in joules)
F is the force applied (in newtons)
d is the distance (in metres)

Example: A force of 70 N moves a wheelbarrow over a distance of 30 m. Calculate the work done.

$E_w = F \times d = 70 \times 30 = 2100$ J (or 2·1 kJ)

Look out for

Double check your energy change answers to see that you do not have your answer the wrong way round – a common mistake.

General question 1

A piano of mass 250 kilograms is pushed up a ramp 3 m long by applying a constant force of 600 newtons as shown.

(a) Calculate the weight of the piano. (2)

(b) What is the minimum force required to lift the piano vertically into the van? (1)

(c) Calculate the work done by pushing the piano up the ramp. (2)

(d) How can the force required to push the piano up the ramp be reduced? (1)

General question 1 – Answer

(a) $W = mg$

      $= 250 \times 10$

      $= 2500$ N

(b) The minimum force to lift the piano will be equal to the weight,

      as lifting it at a steady speed means the forces are balanced.

      Answer is 2500 N.

(c) Work done = $E_w = F \times d = 600 \times 3 = 1800$ J

(d) Put the piano on rollers to reduce friction.

The explanation is not required.

Many candidates will think that the answer to this question is 2501 N – a little bit more than the weight! This is wrong as the piano would then accelerate up as the forces are unbalanced. The minimum force would be used when the piano moves up at a steady speed.

Look out for

When you are asked for **work done** think:
force × distance.

Power

What you should know at **General** **level...**

Power is the **rate** at which **energy is transformed** or more simply the **work done per second**.

$P = \dfrac{E}{t}$, where *P* is the power in **watts** (the same as **joules per second**)

      E is the energy transformed or work done in joules

      t is the time taken in seconds

Example: An electric motor uses 72 kJ of energy in 3 minutes. Calculate the power of the motor.

$P = \dfrac{E}{t} = \dfrac{72\,000}{3 \times 60} = 400$ W

Look out for

kJ must be changed to joules, and the time must be expressed in seconds, not minutes.

Gravitational potential energy

An object **at height *h*** above the Earth's surface has **gravitational potential energy**. Work is done against the force of gravity moving the object upwards and this work is stored as gravitational potential energy.

The relationship is:

$E_p = mgh$

where E_p is the gravitational potential energy in joules

      m is the mass of the object in kilograms

      g is the gravitational field strength (10 N/kg on Earth)

      h is the height above the Earth's surface in metres

continued

What you should know at General level – continued

Example: A 1·8 kg flowerpot is on a window ledge 3·2 m above the ground.

The gravitational potential energy of the flowerpot is:

$E_p = mgh$
$ = 1·8 \times 10 \times 3·2$
$ = 57·6\ J$

The higher the mass is above ground level, the greater is the gravitational potential energy.

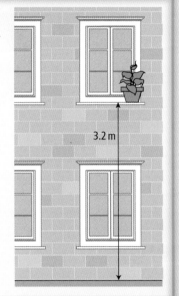

3.2 m

Look out for

The work done lifting the flowerpot into position must be 57·6 J as well. If the flowerpot fell off the window ledge then the stored gravitational potential energy 57·6 J would be converted into kinetic energy.

General question 1

A car is being repaired in a garage. The car is on a ramp and is raised to a height of 1·5 metres. The car has a mass of 1200 kilograms.

(a) Calculate the weight of the car. (2)

(b) Calculate how much gravitational potential energy the car has gained when it is 1·5 metres above the garage floor. (2)

The car is raised in 12 seconds.

(c) Calculate the minimum power needed to lift the car 1·5 metres in 12 seconds. (1)

(d) In practice, the power needed to lift the car in this time is greater than the minimum power. Explain why. (1)

General question 1 – Answer

(a) $W = mg = 1200 \times 10 = 12\ 000\ N$

(b) $E_p = mgh = 1200 \times 10 \times 1·5 = 18\ 000\ J$

(c) $P = \dfrac{E}{t} = \dfrac{18\ 000}{12} = 1500\ W$

(d) Work will have to be done against friction as well as lifting the car

or The platform of the lift has mass and needs to be lifted as well as the car.

Kinetic energy

The **kinetic energy** of a moving object depends on the **mass** and **speed** of the moving object. The relationship for the kinetic energy of a moving object is:

$E_k = \frac{1}{2}mv^2$ where E_k is the kinetic energy in joules (J)
m is the mass of the moving object in kg
v is the speed in m/s

Example: A motorbike of mass 400 kg travelling at 20 m/s has 80 000 J of kinetic energy. Check this calculation. If your answer is 4000 J then you have forgotten to square the speed (a common omission).

Conservation of energy

Gravitational potential energy is always transformed into kinetic energy. The following numerical examples are all at credit level.

Kinetic energy to potential energy $E_k \rightarrow E_p$

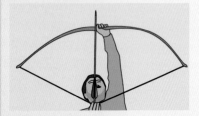

An arrow of mass 0·12 kg is fired vertically into the air with an initial speed of 25 m/s.

The height the arrow reaches can be calculated by equating initial E_k with E_p.

$E_k = \frac{1}{2}mv^2 = 0.5 \times 0.12 \times 25^2 = 37.5$ J

E_k at start = E_p at maximum height h

$E_p = mgh = 37.5$

$h = \dfrac{37.5}{0.12 \times 10} = 31.25$ m

continued

Don't forget to square the speed

Potential energy to kinetic energy $E_p \rightarrow E_k$

A golf ball of mass 0·05 kg is dropped from a height of 15 m. The speed of the ball just as it reaches the ground can be calculated.

$E_p = mgh = 0.05 \times 10 \times 15 = 7.5$ J

E_k at ground = E_p at start = 7·5 J

$\frac{1}{2}mv^2 = 7.5$

$v^2 = \dfrac{7.5}{0.5 \times 0.05} = 300$

$v = \sqrt{300} = 17.3$ m/s

15 m

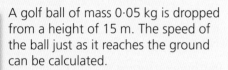
Look out for

Occasionally the mass of the object is not given in a numerical question to calculate v or h.
This is because the mass m cancels when we equate $E_p = E_k$, as follows:

$E_p = E_k$

$\cancel{m}gh = \frac{1}{2}\cancel{m}v^2$ \qquad cancel m

$gh = \frac{1}{2}v^2$

$h = \dfrac{v^2}{2g}$ or $v^2 = 2gh$ which becomes $v = \sqrt{2gh}$

What you should know at Credit level – continued

Potential energy to kinetic energy plus work done against friction $E_p \rightarrow E_k$

Consider a 3 kg box sliding down a slope of length 12 m.

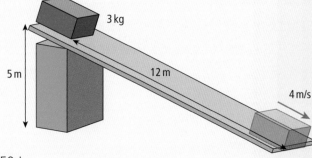

The speed of the box at the bottom of the slope is 4 m/s.

Calculate the force of friction acting on the box as it slides down the slope.

E_p of box at top of slope $= mgh$
$$= 3 \times 10 \times 5 = 150 \text{ J}$$

E_k of box at bottom of slope $= \frac{1}{2}mv^2 = 0{\cdot}5 \times 3 \times 4^2 = 24 \text{ J}$

The work done against friction = the missing energy
$$= 150 - 24$$
$$= 126 \text{ J}$$

$F \times d = 126$

$F \times 12 = 126$

$F = \dfrac{126}{12} = 10{\cdot}5 \text{ N}$

Credit question 1

A skateboarder is practising on a ramp. The total mass of the boarder and the board is 60 kg.

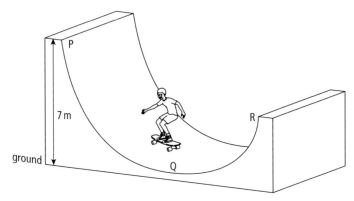

(a) Calculate the increase in gravitational potential energy of the boarder and board in moving from the ground to position P. (2)

(b) The boarder moves along the ramp from P to R and rises into the air above R. At what point **on the ramp** is the kinetic energy of the boarder greatest? (1)

(c) The vertical speed of the boarder at R is 6 m/s. Calculate the height that the boarder rises to above R. (3)

(d) Explain why the boarder does not rise to the same height as P. (2)

Look out for

Part (a) should be an easy 2 marks in the Credit exam. When you see **calculate the gravitational potential energy** think $E_p = mgh$

Credit question 1 – Answer

(a) $E_p = mgh = 60 \times 10 \times 7 = 4200$ J

(b) At point Q.

All the E_p has been turned into E_k at the lowest point.

(c) E_k at point R $= \frac{1}{2}mv^2 = 0.5 \times 60 \times 6^2 = 1080$ J

This 1080 J of E_k will be turned into E_p above point R

$E_p = 1080$ J

$1080 = mgh$

$1080 = 60 \times 10 \times h$ where h is the height above point R

$h = \dfrac{1080}{60 \times 10} = \dfrac{1080}{600} = 1.8$ m

This is a 3 mark question so you will have to do more than one calculation.

(d) Work has been done against friction.

Maximum E_p above point R is less than E_p at point Q

This is an **explain** question, not a **state** question, and is worth 2 marks. There will be two individual marks for two parts to the answer.

Credit question 2

A cyclist is riding an off-road course. The combined mass of the cyclist and bike is 80 kg.

To get to the start of the course the cyclist has pedalled along a slope of 112 m to the top of a hill of height 12·8 km, as shown in the diagram.

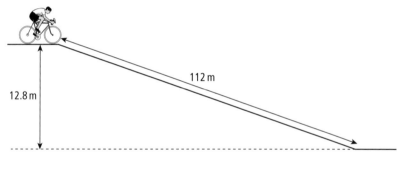

12.8 m

112 m

(a) Calculate how much potential energy has been gained by the cyclist and the bike at the top of the hill. (2)

(b) The cyclist then starts from rest and descends the hill without pedalling, keeping the brakes partly on. There is a constant frictional force of 40 N acting up the slope during the descent. Calculate the amount of work done against friction during the descent. (2)

(c) What is the kinetic energy of the cyclist and bike on reaching the bottom of the hill? (2)

Credit question 2 – Answer

(a) $E_p = mgh = 80 \times 10 \times 12.8 = 10\,240$ J

(b) Work done $= E_w =$ force $\times$ distance $= 40 \times 112 = 4480$ J

(c) E_k at bottom of hill $= E_p$ at top of hill – work done against friction

$\qquad\qquad = 10\,240 - 4480$

$\qquad\qquad = 5760$ J

Be careful with E_p. There is a choice of 2 distances in the diagram. Use the vertical distance when calculating E_p.

The work done against friction will be turned into heat energy. Some of the original Ep will be turned into heat energy and the rest into E_k.

Supply and demand
Sources of energy

What you should know at **General** level...

Fossil fuels (coal, oil and gas) are the main sources of energy at present. However they are also finite which means that the supply will eventually run out. Energy sources can be classified as **renewable** and **non-renewable**. Coal, oil and gas are non-renewable energy sources – they are used up once they have been burned.

Conserving (saving) energy can be achieved in a number of ways:

▶ industry – making machinery more efficient; recycling waste heat

▶ the home – use more energy saving/efficient appliances such as light bulbs; more insulation

▶ transport – use more fuel efficient cars or public transport.

Energy usage is measured in units of joules (J) and kilowatt-hours. Large amounts of energy are measured in kJ, MJ, GJ and TJ.

Look out for

Make sure you know what the prefixes k, M, G and T represent.

Renewable	Non-renewable
Wind	Coal
Tides	Peat
Hydro-electric	Oil
Waves	Gas
Geothermal	uranium
Solar	

General question 1

A small wind turbine costs £1600 to install and is expected to last for 20 years. In 2006 it generated 1·5 kW of electricity for 2000 hours.

(a) Calculate the energy generated in kilowatt-hours during 2006.　(2)

(b) A customer buying this electricity paid 8 pence per kilowatt-hour. Calculate the total cost of buying this electrical energy.　(2)

General question 1 – Answer

(a) kilowatt-hours = kilowatts × hours = 1·5 × 2000 = 3000 kilowatt-hours

(b) cost = kilowatt-hours × cost per kilowatt-hour = 3000 × 8

= 24 000 p (£240)

Notice that the £1600 installation cost is not used in either (a) or (b). You must select the correct costing information to solve this problem.

Look out for

The kilowatt-hour is a unit of **energy** and not power. Many students think it is a unit of power because it contains the word kilowatt.

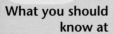

What you should know at Credit **level...**

Renewable energy sources are generally clean (and do not produce CO_2 when generating electricity) and the energy will not run out. There are disadvantages to renewable energy sources, which depend on one particular source and it is often expensive to install the necessary equipment to collect and convert the energies.

Renewable energy source	Disadvantages
Solar	The weather is not always sunny.
Hydroelectric	It needs water available in high areas.
Wind power	Needs a lot of windmills for a reasonable output which can be noisy and an eyesore.
Tidal, wave power and geothermal	These are limited to a few suitable areas because of their nature.

Look out for

There are very few questions in the Credit Level exam on **Supply and demand**. Nearly all appear in the General Level exam.

Generation of electricity

Power stations

What you should know at General **level...**

Electricity generation requires transformation of one type of energy to at least one other type of energy, and in some cases, there are a number of energy transformations. There are different energy transformations for:

▸ thermal electricity generation

▸ hydroelectric generation

▸ nuclear electricity generation.

Thermal power station

In a thermal power station, a fossil fuel (usually coal or gas) containing **chemical energy** is burned to produce **heat energy**. This heat energy boils water and the steam produced turns the blades of a turbine. The energy change here is **heat energy** to **kinetic energy.** The rotating turbine is connected to a generator which produces electricity. The energy change in the generator is **kinetic energy** to **electric energy**.

Hydroelectric power station

The hydroelectric power station uses fast moving water to turn the generator.

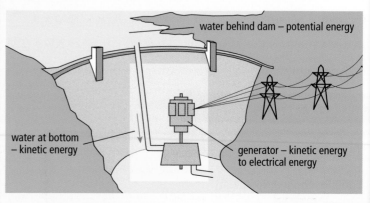

water behind dam – potential energy

water at bottom – kinetic energy

generator – kinetic energy to electrical energy

continued

Pumped hydroelectric schemes

A pumped hydroelectric power station pumps water back up to the reservoir using cheaper off-peak electricity at night when there is low demand. This water flows back down to the generator during daytime generating electricity which is sold to customers at the peak (higher) rate.

Nuclear power station

The nuclear power station uses nuclear fuel to produce **heat energy** in the nuclear reactor.

Steam is then used to turn the generator. The complete energy changes are shown in the diagram:

Nuclear energy → **Heat energy** → **Kinetic energy** → **Electrical energy**

What you should know at **Credit** level...

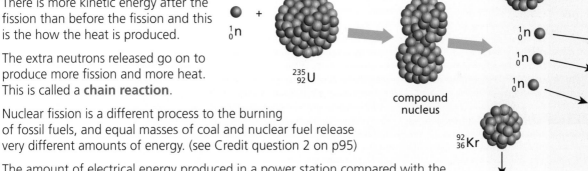

In nuclear reactors, a uranium nucleus is bombarded by a neutron and splits into two smaller nuclei and more neutrons. This splitting is called **fission**. There is more kinetic energy after the fission than before the fission and this is the how the heat is produced.

The extra neutrons released go on to produce more fission and more heat. This is called a **chain reaction**.

Nuclear fission is a different process to the burning of fossil fuels, and equal masses of coal and nuclear fuel release very different amounts of energy. (see Credit question 2 on p95)

The amount of electrical energy produced in a power station compared with the energy released by the fuel is a measure of the **efficiency** of the generation process.

General question 1

Coal is a fossil fuel that is used to generate electricity.

In a coal-fired power station, identify the energy transformation in the boiler, turbine and the generator.

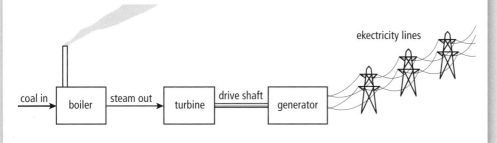

General question 1 – Answer

Boiler: **chemical** to **heat**

Turbine: **heat** to **kinetic**

Generator: **kinetic** to **electrical**

Look out for

A common mistake students make is writing the energy change the wrong way round. Always double check energy change answers to confirm the correct order has been given.

General question 2

In a pumped hydroelectric scheme, water is stored in a reservoir 400 metres above the power station.

(a) Describe what is meant by a **pumped** hydroelectric scheme. (2)

(b) Give **one** advantage of a pumped hydroelectric scheme. (1)

(c) Calculate the potential energy transferred by one kilogram of water in moving from the reservoir to the power station. (2)

General question 2 – Answer

(a) Water is pumped back up to the reservoir using excess electricity
     at night.

(b) reservoir never runs out of water

(c) E_p = mgh = 1 × 10 × 400 = 4000 J

This is a two mark question so two different statements are required for full marks.

Calculating gravitational potential energy appears in every General exam.

General question 3

Some of the stages in a nuclear power station are shown:

| Reactor | → | Turbine | → | Generator | → | National grid |

At what stage is the main energy transformation:

(A) kinetic to electrical; (1)

(B) nuclear to heat? (1)

General question 3 – Answer

(A) generator

(B) reactor

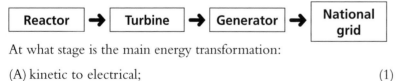

Don't confuse the generator with the turbine. The turbine does not produce electricity.

Look out for

Radioactive waste is produced from nuclear reactors which has to be stored for many years until it has decayed to a safe level. See Chapter 3 Health Physics for more information.

Credit question 1

In the reactor of a nuclear power station, neutrons split uranium nuclei to produce heat in a chain reaction. Explain what is meant by the term **chain reaction**. (2)

Credit question 1 – Answer

More neutrons are released (in each fission). These neutrons cause more fissions.

Note: two marks allocated so make sure you have two parts to the answer.

Credit question 2

In a coal-fired power station 1 kg of coal produces 30 MJ of heat energy.

In a nuclear power station 1 kg of uranium fuel produces 4.5×10^{12} J of heat.

Calculate how many kilograms of coal are required to produce the same amount of heat energy as 1 kg of uranium. (1)

Notice that 1 kg of uranium produces approximately 1 million times more heat that 1 kg of coal.

Credit question 2 – Answer

$$\text{amount of coal} = \frac{4.5 \times 10^{12}}{30 \times 10^{6}} = 150\,000 \text{ kg } (1.5 \times 10^{5} \text{ kg})$$

This is a problem solving question which has no formula.

Efficiency of a power station

What you should know at **Credit** **level...**

The efficiency of a power station (or any machine) is given by:

$$\text{Percentage efficiency} = \frac{\text{useful energy out}}{\text{total energy in}} \times 100$$

$$\text{Percentage efficiency} = \frac{\text{useful power out}}{\text{total power in}} \times 100$$

Look out for

Percentage efficiency does not have any units. The % sign may be included.

Credit question 1

A power station produces 500 MW of heat energy
The generator connected to the turbine produces 350 MJ of electrical energy.
Calculate the efficiency of this generator. (2)

Credit question 1 – Answer

$$\text{Percentage efficiency} = \frac{\text{useful energy out}}{\text{total energy in}} \times 100$$

$$= \frac{350 \times 10^{6}}{500 \times 10^{6}} \times 100$$

$$= 70\%$$

Source to consumer

Generation of alternating current

What you should know at **General** **level...**

It is possible to create a voltage and current in a coil of wire if a magnet is moved towards (or away from) the coil.

Alternatively a current can be generated if the coil is moving and the magnet is stationary. The main parts of an a.c. generator are:

▶ coil

▶ iron core

▶ magnet.

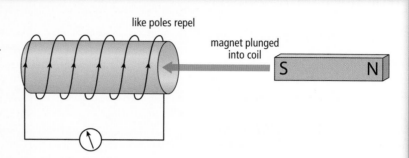

What you should know at **Credit** **level...**

The factors which affect the size of the induced voltage in a generator are:

▶ the strength of the magnet or magnetic field

▶ the number of turns on the coil

▶ the relative speed between the magnet and the coil.

The greater these are, the greater the size of the induced voltage.

A changing magnetic field induces a voltage in a coil.

In full-size generators used in power stations, electromagnets are used in place of permanent magnets.

General question 1

In which of the following would a voltage **not** be induced in a coil of wire? (1)

(A) Rotating the coil of wire near a magnet.

(B) Rotating a magnet near to the coil of wire.

(C) Holding a magnet stationary within the coil of wire.

(D) Moving a magnet in and out of the coil of wire.

(E) Moving the coil of wire between the poles of a magnet.

General question 1 – Answer

(C)

All the others have some relative movement between the coil and the magnet.

General question 2

The diagram shows how a bicycle dynamo is constructed.

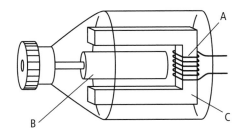

Use the names given below to identify the three main parts of the dynamo and match them with the labels on the diagram. (2)

coil   iron core   magnet

General question 2 – Answer

A _coil_   B _magnet_   C _iron core_

The moving part in this dynamo (generator) must be the magnet

Credit question 1

A power station uses an a.c. generator to convert kinetic energy from a turbine into electrical energy, as shown:

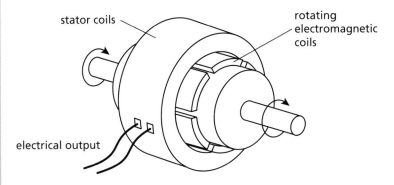

(a) Explain how the a.c. generator works. (2)

(b) State **two** changes that can be made to the generator to increase the output power. (1)

Credit question 1 – Answer

(a) _The rotating electromagnetic coils produce a changing magnetic field. This induces a voltage (or current) in the stator coils._

The word 'changing' is essential in this answer.

(b) **1** _increase the speed of rotation_

   2 _increase the number of stator coils (or rotating coils)_

Transformers

What you should know at **General** level...

Transformers are used to change the size of an a.c. voltage.

Transformers have an input coil called a **primary coil** and an output coil called a **secondary coil**. An iron core links both coils. The input and output voltages are related using this expression:

$$\frac{V_s}{V_p} = \frac{n_s}{n_p}$$

What you should know at **Credit** level...

Transformers are not 100% efficient because of energy losses in the coils caused by heating of the wires. Also a small amount of energy is lost as sound energy when the transformer makes a distinctive humming sound.

The relationships $\frac{V_s}{V_p} = \frac{n_s}{n_p}$ and $\frac{I_p}{I_s} = \frac{n_s}{n_p}$ are used to solve numerical calculations.

Note the current formula has the I_s and I_p subscripts the other way round from the voltage formula. Care is required when using the current formula (relationship).

The ratio of turns in the secondary coil to turns in the primary coil is called the **turns ratio**.

General question 1

A 5 volt battery in a mobile phone is recharged from the mains using a charger containing a step-down transformer. The transformer consists of three parts:

   core   **primary coil**
secondary coil

(a) Show these parts on the diagram. (3)

core

230 volt mains

5 volts

primary coil   *secondary coil*

(b) There are 11 500 turns on the primary coil of the transformer. Calculate the number of turns on the secondary coil. (2)

$$\frac{V_s}{V_p} = \frac{n_s}{n_p}$$

$$\frac{5}{230} = \frac{n_s}{n_p} = \frac{n_s}{11\,500}$$

$$n_s = \frac{5 \times 11\,500}{230} = 250 \text{ turns}$$

(c) **Explain** why a transformer cannot be used to step down the voltage from a battery. (2)

Transformers only operate on a.c., a battery supplies d.c.

Look out for

Using $\frac{V_s}{V_p} = \frac{n_s}{n_p}$ appears in every General level exam

Look out for

Take care with rearranging the second line. Make sure you get plenty of practice at rearranging transformer data to get the final answer.

Look out for

Remember d.c. does **not** produce a **changing** magnetic field essential to the working of a transformer.

Credit question 1

A power station generates an electric current of 5000 A at 20 kV. The output from this generator is connected to a transformer which steps up the voltage to 132 kV.

There are 2000 turns in the primary circuit of the transformer. Assuming the transformer is 100% efficient:

(a) calculate the number of turns in the secondary coil; (2)

(b) calculate the current in the secondary coil of the transformer. (2)

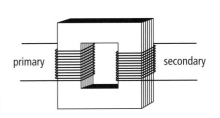

primary   secondary

Credit question 1 – Answer

(a) $\dfrac{V_s}{V_p} = \dfrac{n_s}{n_p}$

$\dfrac{132\ 000}{20\ 000} = \dfrac{n_s}{2000}$

$n_s = \dfrac{132\ 000 \times 2000}{20\ 000} = 13\ 200$ turns

(b) $\dfrac{I_p}{I_s} = \dfrac{n_s}{n_p}$

$\dfrac{5000}{I_s} = \dfrac{13\ 200}{2000}$

$I_s = \dfrac{5000 \times 2000}{13\ 200} = 758$ A

or

Power in primary coil = Power in secondary coil
(as transformer is 100% efficient)

$V_p I_p = V_s I_s$

$20\ 000 \times 5000 = 13\ 200\ I_s$

$I_s = 758$ A

Look out for

Remember in the transformer current formula, the subscript $_s$ is **not** on both upper parts of the formula, unlike the transformer voltage formula.

Credit question 2

A battery charger with an input voltage of 230 V is used to recharge a car battery. The charger contains a transformer that has an output voltage of 13·8 V.

(a) What type of transformer does the battery charger contain? (1)

(b) There are 4000 turns in the primary coil of the transformer. Assuming the transformer is 100% efficient, calculate the number of turns in the secondary coil. (2)

(c) When charging the battery, the current in the secondary coil is 4.7 A. Calculate the power output of the transformer. (2)

(d) In practice, the transformer is only 94% efficient. Calculate the current in the primary coil. (3)

(e) State and explain **one** reason why a transformer is not 100% efficient. (2)

Credit question 2 – Answer

(a) step-down transformer

(b) $\dfrac{V_s}{V_p} = \dfrac{n_s}{n_p}$

$\dfrac{13 \cdot 8}{230} = \dfrac{n_s}{4000}$

$n_s = \dfrac{13 \cdot 8 \times 4000}{230} = 240$ turns

(c) $P_o = IV$

$= 4 \cdot 7 \times 13 \cdot 8$

$= 64 \cdot 9$ W

(d) (new) power output = 94% of 64·9 = 0·94 × 64·9 = 61 W

$I_o V_o$ = Power output = 61 W

$I_o \times 13 \cdot 8 = 61$

$I_o = 4 \cdot 4$ A

(e) any one from
- power loss due to heating in the coil
- power loss due to sound generated in the transformer
- power loss due to energy required to magnetise the iron core
- power loss due to heating in the wires due to a combination of current and resistance

*230 V to 13·8 V so **voltage** going **down***

This is a three-mark question, so there is more than one calculation

Credit question 3

A pupil investigates the efficiency of transformers using a transformer, two joulemeters and a lamp connected to a supply as shown:

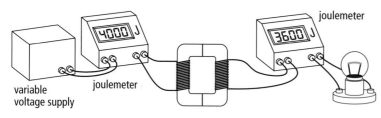

Initially, the displays on both joulemeters are set to zero. The supply is switched on and after a certain time the readings shown are recorded. Calculate the percentage efficiency of the transformer.

Credit question 3 – Answer

efficiency $= \dfrac{\text{energy output}}{\text{energy input}} \times 100$

$= \dfrac{3600}{4000} \times 100 = 90\%$

Remember the output is at the end, and the input at the start.

National Grid

What you should know at General **level...**

The National Grid is a series of transmission lines (or electrical cables) which carry electricity across the country. These transmission lines are carried on pylons.

High voltages are used in the transmission of electricity to reduce power loss.

Stepping up the voltage reduces the current and it is the current which produces heat energy. Less current in the cable means less heating of the cable.

What you should know at Credit **level...**

Electricity has to be transferred from power stations to the consumer via the **National Grid**. The electricity has to be sent for many kilometres through the overhead cables strung from pylons. The length of these transmission lines means that energy losses due to the heating effect of the current flowing through the cables could be very high. The power loss, P, in watts (W) is given as:

$P_{loss} = I^2R$, where: I = current through transmission line (A)
R = total resistance of the transmission line (Ω)

At the power station a step-up transformer is used to increase the voltage.

This reduces the current in the transmission line cables and so reduces the power loss as power loss = I^2R.

General question 1

Complete the following passage:

In the National Grid, _step-up_ transformers are used to increase the 25 000 volts from a power station to 132 000 volts for transmission.

This reduces _power loss_ in the transmission lines.

The voltage is then decreased to 11 000 volts for industry and 230 volts for domestic use using _step-down_ transformers.

Credit question 1

Electricity is transmitted along a section of the National Grid. The current in the transmission line is 760 A while the voltage across the cable is 132 kV. The cable has a length of 220 km and a resistance of 0·31 Ω/km.

Calculate the power loss in the cable. (3)

Credit question 1 – Answer

Total resistance of the cable = 220 × 0·31

= 68·2 Ω

$P = I^2R = 760^2 \times 68·2 = 3·94 \times 10^7$ W (or 39·4 MW)

This is a three-mark question, so there is more than one calculation

Heat in the home

Energy conservation in buildings

What you should know at **General** **level...**

The heat loss from a building depends on the temperature **difference** between the inside and outside of a house. There will be more heat loss from a house when the outside temperature is 0°C than when the outside temperature is 5°C.

Heat is lost from the home due to **conduction**, **convection** and **radiation**. It is important to remember that heat is **different** from temperature.

General question 1

Which of the following is a unit of heat? (1)

(A) degree Celsius

(B) joule

(C) joule per kilogram

(D) joule per kilogram per degree Celsius

(E) watt

Look out for

Heat is a form of energy and energy is measured in joules. Temperature is a measure of how hot or cold something is and the unit of temperature is degree Celsius (°C).

General question 1 – Answer

B

General question 2

Heat can be lost from the home in three ways – conduction, convection and radiation.

For each of these ways, state a method by which heat loss can be reduced.

You must give three **different** methods of reducing heat loss. (3)

Look out for

Convection describes warm air rising up taking heat energy with it. Convection currents occur in liquids and gases where the atoms are free to move. Conduction occurs in solids where the atoms vibrate against each other. Radiation does not require atoms to transfer heat. This is how heat from the **sun** reaches Earth.

General question 2 – Answer

Conduction: any one from:
- *loft insulation* • *wall insulation* • *carpets* • *double glazing*

Convection: any one from:
- *loft insulation* • *draft excluders* • *(under) floor insulation*

Radiation: any one from
- *foil behind radiators* • *(closed) curtains* • *foil backed plasterboard*

General question 3

A house is designed to conserve as much energy as possible. The temperature in the house is kept at a constant value while the temperature outside changes. The graph shows the temperature inside and outside the house over a 24 hour period.

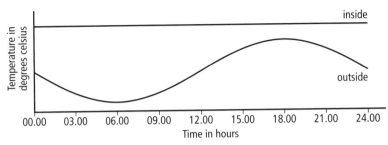

Write down the time at which the heat loss from the house is greatest.

General question 3 – Answer

06·00am because this is when there is the greatest difference between outside and inside temperatures.

Specific heat capacity

What you should know at **General** **and** **Credit** **level…**

The same mass of different materials requires different quantities of energy to raise the temperature of the mass by one degree. 1 kg of water will heat up at a different rate than 1 kg of copper.

The energy required to increase or decrease the temperature of a substance is given by:

$E_h = cm\Delta T$, where: E_h = energy required (J)
c = specific heat capacity of the substance (J/kg°C)
m = mass of the substance (kg)
ΔT = change in temperature of the substance (°C)

The Δ symbol (pronounced **delta**) is used to represent a **change** in some quantity. For example, if an object's temperature were to increase from 17°C to 22°C, ΔT would be 5°C.

When calculating the heat energy in specific heat capacity questions, it may be necessary to use one of a number of expressions for energy:

▶ $E = P \times t$

▶ $E = (VI)t$

▶ $E = F \times d$

▶ $E = \frac{1}{2}mv^2$

Look out for

The value of C, the **specific heat capacity** of a substance, will be given in General level questions, as there is no data sheet at the start of the General Level exam paper.

General question 1

A deep fat fryer contains 1·2 kilograms of liquid cooking oil at a temperature of 20 degrees Celsius. The oil has to be heated to a cooking temperature of 180 degrees Celsius.

(a) Calculate the energy required to raise the temperature of the oil from 20 degrees Celsius to the cooking temperature. (The specific heat capacity of the cooking oil is 2200 joules per kilogram per degree Celsius.) (2)

(b) The heater in the fryer is rated at 2400 watts. Calculate the minimum time for the oil to reach the cooking temperature. (2)

(c) In practice, the actual time is greater than the minimum time calculated. Give one reason for this difference. (1)

General question 1 – Answer

(a) $E_h = Cm\Delta T$

$\quad = 2200 \times 1\cdot2 \times (180 - 20)$

$\quad = 4\cdot2 \times 10^5 \ J$

(b) $\quad E = P \times t$

$\quad 4\cdot2 \times 10^5 = 2400 \times t$

$\quad t = \dfrac{4\cdot2 \times 10^5}{2400} = 175 \ s$

(c) some heat is lost to the surroundings

$E_h = Cm\Delta T$ comes up each year in the General and Credit exam.

ΔT has to be calculated separately. ΔT is not 180°C or 20°C.

Look out for

Part (c) is a very common question and answer. Don't forget the **to the surroundings** part.

Credit question 1

A student sets up the apparatus shown to measure the specific heat capacity of an aluminium block.

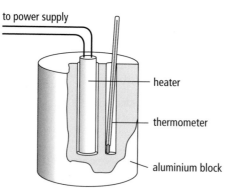

to power supply

heater

thermometer

aluminium block

The student obtains the following results:

mass of aluminium block $m = 0\cdot8$ kg

temperature change $\Delta T = 19$°C

time taken $t = 5\cdot0$ minutes

heater current $I = 4\cdot2$ A`

heater voltage $V = 12$V

(a) Show, by calculation, that 15 120 J of electrical energy are supplied to the heater in 5·0 minutes. (2)

(b) Assuming all of the electrical energy is transferred to the aluminium block as heat energy, calculate the value of the specific heat capacity of aluminium from this experiment. (2)

(c) The accepted value of the specific heat capacity of aluminium is 902 J/kg°C. Give a reason for the difference between your answer in part (b) and this value. (1)

(d) How could the experiment be improved to reduce this difference? (1)

Credit question 1 – Answer

(a) $E = Pt$

$= (VI)t$

$= 12 \times 4·2 \times (5 \times 60)$

$= 15\ 120\ J$

(b) $E_h = Cm\Delta T$

$15\ 120 = C \times 0·8 \times 19$

$C = \dfrac{15\ 120}{(0·8 \times 19)} = 995\ J/kg°C$

Take care using a calculator to put brackets around the $0·8 \times 19$ otherwise your calculated answer will be incorrect.

(c) Heat energy has been lost to the surroundings. The heat energy heating up the aluminium only will be less than 15 120 J.

(d) Insulate the aluminium block to reduce heat loss to the surroundings.

Remember **to the surroundings** must be included in your answer.

Change of state

What you should know at General level...

There are three states of matter – **solid**, **liquid** and **gas**.

The change of state from solid to liquid is called **melting**, and liquid to solid is called **solidifying** or **freezing**.

The change of state from liquid to gas is called **vaporisation**, and gas to liquid is called **condensation**.

There is **no change in temperature** when a change of state takes place. For example, when water at 100°C boils to steam at 100°C, the temperature doesn't change.

Energy is **gained or lost** when a change of state takes place. For example, water at 100°C needs energy to change into steam at 100°C even although there is no increase in temperature.

The energy needed for 1 kg of a substance to change state from solid to liquid is called the **specific latent heat of fusion** of the substance.

The energy needed for 1 kg of a substance to change state from liquid to gas is called the **specific latent heat of vaporisation** of the substance.

What you should know at **Credit** **level...**

E_h is the heat energy required to change the state of a mass m of substance at its melting point (using $E_h = mL_f$) or boiling point (using $E_h = mL_v$)

L_f is the **specific latent heat of fusion** of a substance. L_v is the **specific latent heat of vaporisation** of a substance. A list of values of the specific latent heats of fusion and vaporisation of materials is given on the data sheet at the start of the Credit Level exam.

There is no need for ΔT in these formulae as the temperature does not change.

Look out for

Be careful not to extract information for L_f of water from the wrong list. Fusion means melting. (Think – a fuse melts)

General question 1

A cool-box is used to keep food cold for a picnic. The box is well insulated and has an insulated lid. Before food is put in the cool-box, a coolant pack is placed in a freezer. The coolant in the pack changes from liquid to solid. The coolant pack is then placed in the cool-box with the food.

The temperature changes of the food and the coolant over time are shown in the graph.

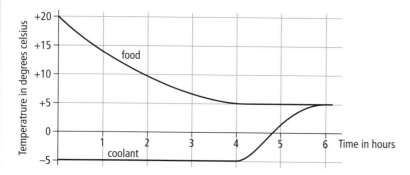

(a) What is the final temperature of the food? (1)

During the first 4 hours, the temperature of the coolant does not change but the temperature of the food falls.

(b) What happens to the coolant in this time? (1)

(c) Explain why the temperature of the food falls during this time. (2)

(d) Why is the box insulated? (1)

General question 1 – Answer

(a) +5°C

(b) The coolant is melting

(c) Because the coolant is melting it takes energy from the food

(d) So less heat from the outside can heat up the food inside the cool-box.

The temperature stays constant as the melting takes place.

This is a 2 mark question. So marks are allocated as follows Coolant takes in heat (1) from the food (1)

*This is the reverse of the very common answer **heat loss to the surroundings.***

General question 2

A coolant pack is used to treat an injured player at a hockey match. Before the match the coolant pack is stored in a refrigerator at 2°C. The coolant inside the pack changes from liquid to solid. The coolant has a melting point of 7°C and a mass of 0·5 kg. The coolant pack is removed from the refrigerator and placed on the injured ankle of the player.

(a) Calculate the energy required to raise the temperature of the coolant pack from 2°C to its melting point. (Specific heat capacity of coolant = 2100 J/kg°C.)   (3)

(b) Where does most of the energy required to raise the temperature of the coolant come from?   (1)

(c) Having reached its melting point, the coolant pack then remains at the same temperature for 15 minutes. What happens to the coolant in this time?   (1)

(d) One of the other players suggests insulating the coolant pack and ankle with a towel. Why should this be done?   (1)

General question 2 – Answer

(a) $\Delta T = 7 - 2 = 5°C$

$E_h = Cm\Delta T$

$\quad = 2100 \times 0.5 \times 5$

$\quad = 5250 \text{ J}$

(b) player's ankle

(c) coolant is changing state (or melting)

(d) to reduce heat transfer from the surroundings to the coolant pack.

Credit question 1

A mass of 500 g of a substance is heated with a 30 W heater. A temperature probe is inserted into the substance.

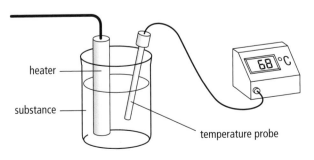

107

The substance is initially solid and at room temperature. The graph shows the variation of the temperature of the substance from the time the heater is switched on.

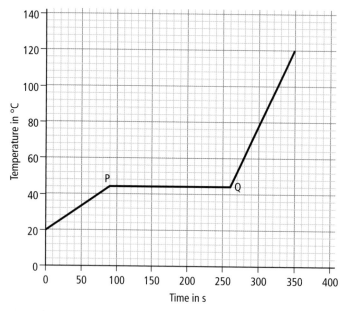

(a) State the value of room temperature. (1)

(b) Why does the temperature of the substance remain constant between P and Q? (1)

(c) Calculate the energy transferred by the heater during the time interval PQ. (3)

(d) Calculate the specific latent heat of fusion of the substance. (1)

Credit question 1 – Answer

(a) 20°C

(b) The solid is melting/changing state.

(c) Time between P and Q = 260 – 90 = 170 s

$E = P \times t$

$= 30 \times 170$

$= 5100$ J

(d) $E_h = mL_f$

$5100 = 0.5 \times L_f$

$L_f = \dfrac{5100}{0.5}$

$= 10\,200$ J/kg

Signals from space

Astronomical terms

Term	Definition
Planet	A natural satellite which orbits a sun.
Moon	A natural satellite which orbits a planet.
Star	A mass of gas which emits heat and light.
Solar System	A star and the objects which surround it.
Sun	The star at the centre of our solar system.
Galaxy	A collection of stars. Our galaxy is called the Milky Way.
Universe	Consists of many millions of galaxies separated by space.

The **light year** is a unit of **distance** used in astronomy. A **light year** is defined as the distance which light travels in one year.

$d = v\,t$ 1 light year = $3 \times 10^8 \times 365 \times 24 \times 60 \times 60$
= $9{\cdot}46 \times 10^{15}$ m

To reach the Earth in terms of time, light takes approximately:

▶ 8 light minutes from the Sun,

▶ 4·3 light years from the next nearest star

▶ 100 000 light years from the edge of our galaxy.

Look out for

Think! Why are metres (or kilometres) not suitable as an astronomical unit of distance?

The refracting telescope

The main features of a **refracting** telescope (i.e. one containing glass lenses) are two convex lenses known as the **objective lens** and the **eyepiece lens** contained in a light–tight tube.

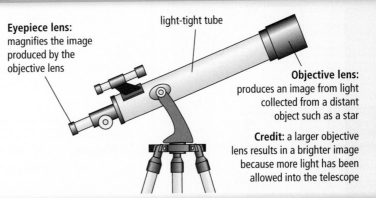

Eyepiece lens: magnifies the image produced by the objective lens

light-tight tube

Objective lens: produces an image from light collected from a distant object such as a star

Credit: a larger objective lens results in a brighter image because more light has been allowed into the telescope

General question 1

The diagram below shows a refracting telescope, which is used by astronomers to view distant stars, planets and galaxies.

(a) Which lens, the objective or the eyepiece, has the longer focal length?

(b) What is the purpose of the eyepiece lens?

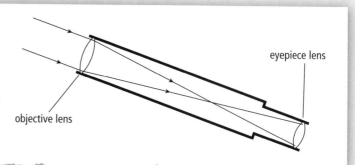

General question 1 – Answer

(a) *The objective lens*

(b) *to magnify the image*

The eyepiece has a shorter focal length because it provides the magnification. The objective lens is "thinner" as it has a longer focal length.

Ray diagrams

What you should know at **Credit** **level...**

Ray diagrams are used to work out the position of an image formed by a convex lens. Two rules are used to determine the position of an image:

▶ **Rule 1** When a light ray from the top of an object enters the *centre* of the lens, it passes straight through. (1)

▶ **Rule 2** When a horizontal light ray from the top of the object enters the lens it is refracted towards the focus of the lens. (2) The image is formed where these 2 rays meet.

Someone looking at A sees an inverted (upside down) **real image** which can be focused on a screen.

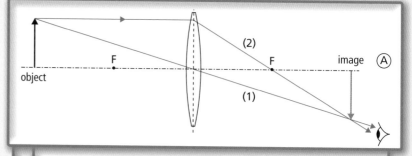

Magnifying glass

When an object is placed between a convex lens and the focus, the convex lens acts as a magnifying glass.

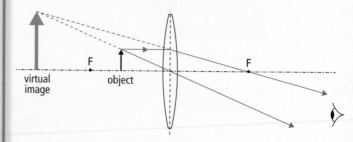

Applying the two rules as above, the rays only meet when projected back as shown by the dotted lines. The human eye focuses on this virtual image which cannot be focused on a screen – just like when you look at your virtual image in a mirror.

Look out for

Make sure you can draw these diagrams (using a ruler) for Credit questions.

Credit question 1

An astronomer examines a photograph using a magnifying glass. Complete a ray diagram to show how the magnifying glass can be used to form an image of the photograph. Your diagram must show the position of the image. (3)

Credit question 1 – Answer

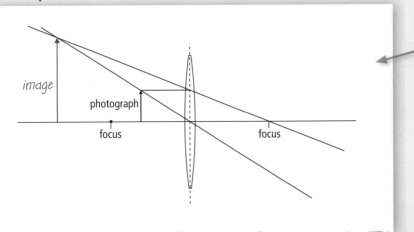

Solid lines will do. It is not necessary to have the virtual parts of the lines dotted. This ray diagram question appears every two or three years and is usually not done so well by students. Take some time to practice this question.

Refraction of light and spectroscopy

What you should know at **General** **level...**

When white light passes through a triangular glass prism it separates into its different colours corresponding to the different wavelengths. The glass prism is a simple spectroscope used to study wavelengths in light.

The prism **refracts** each colour by a different amount, with violet light (shortest wavelength) being the colour which is refracted most and red light refracted least (longest wavelength).

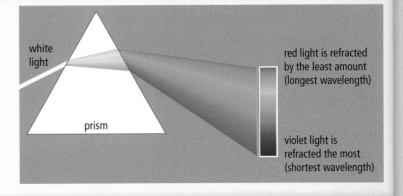

Light from a star can be studied using a **spectroscope** containing a prism which separates the light from the star into the various different colours (or wavelengths) which it contains.

Each element emits a characteristic spectrum composed of light of particular wavelengths. This is called the element's **line emission spectrum** and the elements that make up a star can be identified by the lines emitted. The diagram below shows the line emission spectrum for sodium.

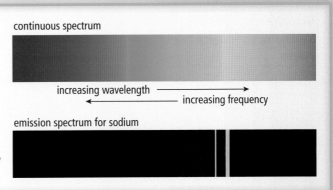

The electromagnetic spectrum

What you should know at **General** level...

From earlier work we have learnt that visible light is only a very small part of the whole electromagnetic spectrum.

The electromagnetic spectrum is a large family of waves with a wide range of wavelengths which all travel at the speed of light (3×10^8 m/s). However they each have different wavelengths, frequencies and energies.

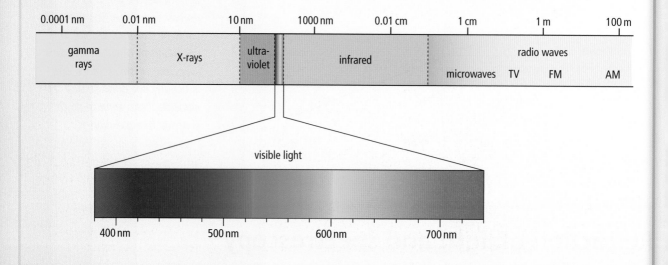

What you should know at **Credit** level...

Learn the order of the waves in terms of wavelength and frequency and give an example of a detector for each.

Radiation	Suitable detector	Frequency	Wavelength
Gamma rays	Geiger counter	high	short
X-rays	Photographic film		
Ultra-violet	Fluorescent chemicals		
Visible light	Human eye and digital photo cameras		
Infra-red	Infra-red cameras thermometer		
Microwaves	Aerial		
TV and radio waves	Aerial	low	long

Look out for

You do not need to know their exact wavelengths – simply the order in which they are placed!

Look out for

Remind yourself why these waves are studied in Telecommunications and Health Physics.

Telescopes

What you should know at **General** and **Credit** level...

As well as the optical telescope, with which everyone is familiar, other types of telescope also exist. Astronomers can use the entire electromagnetic spectrum to study radiations from outer space. However, they must use different detectors or instruments to pick up the different types of waves

Telescopes can be designed to detect radio waves using curved reflectors.

Radio waves from space are reflected onto the aerial or antenna placed at the focus of the telescope

This telescope is made of metal to absorb the radio waves. Remind yourself of the purpose of an aerial in radio reception in Chapter 1.

Look out for

What is the purpose of the curved dish? Look at curved reflectors in Chapter 1.

Credit question 1

A team of astronomers observes a star 200 light years away. Images of the star are taken with three different types of telescope as shown.

(a) Explain why different types of telescope are used to detect signals from space.

(b) Place the telescopes in order of the increasing wavelength of the radiation which they detect.

(c) State a detector that could be used in telescope C.

Telescope A – visible light

Telescope B – infra-red

Telescope C – X-ray

Credit question 1 – Answer

(a) *different detectors are required for different radiations/ frequencies/wavelengths*

(b) *C A B*

(c) *G-M tube **or** photographic film*

7

Space travel
Rocket propulsion

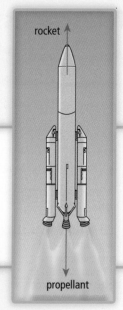

rocket

propellant

What you should know at General **level...**

Rockets are pushed **forward** because the propellant gas is pushed **backwards**. The acceleration of the rocket can be calculated using Newton's Second Law $F = ma$ which was met previously in Chapter 5. To calculate the acceleration, the mass and force (engine thrust) of the rocket must be known.

The rocket will move at a **steady speed** once the rocket **motor is switched off** in deep space. The forces are balanced (zero forces acting on the rocket) so by Newton's First Law the rocket moves at a steady speed.

What you should know at Credit **level...**

The motion of a rocket can be explained by Newton's Third Law which states:

If A exerts a force on B then B exerts an **equal and opposite** force on A.

The rocket forces the propellant backwards. By Newton's Third Law the rocket propellant exerts an equal and opposite force on the rocket causing the rocket to move forward.

Equal and opposite forces which illustrate Newton's Third Law are called **Newton pairs.**

General question 1

A spacecraft is far out in space. An astronaut wearing a backpack leaves the spacecraft. He uses the backpack to move around. The backpack contains a pressurised gas cylinder connected to a valve. When the valve is opened, a jet of gas is released.

(a) Select the correct word in the following passage:
When the astronaut opens the valve, the cyclinder pushes gas backwards. The gas pushes the cylinder/jet/spacecraft forewards. (1)

(b) The astronaut and the backpack have a combined mass of 120 kg. The jet of gas exerts a constant thrust of 24 newtons. Calculate the acceleration of the astronaut when the jet is switched on. (2)

(c) Describe and explain the motion of the astronaut when the jet is switched off. (2)

General question 1 – Answer

(a) cylinder

Reverse the wording: **cylinder pushes gas** so **gas pushes cylinder**

(b) $F = ma$

$24 = 120 \times a$

$a = \dfrac{24}{120} = 0.2 \text{ m/s}^2$

This is an easy 2 mark calculation on Newton's Second Law $F = ma$ but remember the unit is m/s².

(c) The astronaut travels at a steady speed because there is no unbalanced force. (or the forces on the astronaut are balanced.)

This is a **Describe and explain** question worth 2 marks, so there are two parts to the answer.

Credit question 1

Explain in terms of Newton's Third Law how a rocket is able to move forward. (2)

> Your answer should include mention of a Newton pair of forces.

Credit question 1 – Answer

The rocket exerts a force on the gas. The gas exerts an equal and opposite force on the rocket.

! Look out for

To describe a Newton pair of forces simply reverse the wording of the first force to describe the second force.

Gravity and weightlessness

What you should know at **General** **level...**

Weight and the force of gravity have been studied in Chapter 5. In space physics the acceleration due to gravity is studied in detail.

All objects have the **same acceleration** when falling close to the Earth's surface, if air resistance is negligible.

The acceleration due to gravity on Earth is 10 m/s^2. If a £2 coin and a 1 p coin are dropped simultaneously from the same height they will both have an acceleration of 10 m/s^2 and hit the ground at the same time.

Objects in **free fall** appear **weightless**.

The spring balance shows the weight of the stone

Freely falling spring balance with the stone shows a zero reading

General question 1

A fish of mass 2 kilograms is hung on a Newton balance. The fish and the balance are dropped and fall freely to the sea below. What is the reading on the Newton balance **while falling**?

A 0 newton

B 1 newton

C 2 newtons

D 10 newtons

E 20 newtons.

> *Answer A is correct as falling objects appear weightless so reading on scales will be zero.*

General question 1 – Answer

A

General question 2

Two objects are dropped from the same height. Both objects fall freely. Object X has a mass of 10 kg. Object Y has a mass of 1 kg. Object X accelerates at 10 metres per second per second. The acceleration of object Y, in metres per second per second, is:

A 0.1

B 1.0

C 10

D 100

E 1000

> *Answer C is correct as all objects fall to the Earth with the same acceleration due to gravity (10 m/s^2).*

General question 2 – Answer

C

Gravity and acceleration

The acceleration due to gravity on a planet has the **same numerical value** as the gravitational field strength on the planet.

The proof of the last statement has to be understood at Credit Level: the weight of an object of mass m is:

weight = mg

but weight = ma ← ● *Since weight is the unbalanced force downwards,*

∴ $\not{m}a = \not{m}g$ ← ● *cancel m on both sides.*

∴ $a = g$

You should be able to calculate the weight of an object at different places in the Solar System with gravitational field strengths different from 10 N/kg (on Earth). A table of gravitational field strengths is given in the data sheet at the start of the Credit exam paper.

Gravitational field strengths

	Gravitational field strength on the surface in N/kg
Earth	10
Jupiter	26
Mars	4
Mercury	4
Moon	1·6
Neptune	12
Saturn	11
Sun	270
Venus	9

The weight of an object decreases as it moves away from the Earth as shown by this graph.

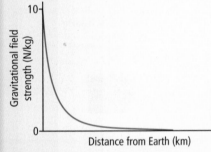

The force of gravity **decreases** as an object moves **further away** from the Earth.

The force of gravity is **zero** in deep space at distances **well away** from the Earth and other planets.

Inertia

Inertia is the name given to an object's resistance to change in speed.

A full supermarket trolley has more inertia than an empty trolley as it is harder to start moving and harder to stop moving.

large inertia

smaller inertia

Credit question 1

(a) Calculate the weight of a 72 kg astronaut on Mars. (2)

(b) A toolbox has a weight of 24 N on the Moon. Calculate the mass of the toolbox. (2)

(c) On which planet would a dropped ball accelerate at 12 m/s²? (1)

Credit question 1 – Answer

(a) $W = mg = 72 \times 4 = 288$ N

(b) $W = mg$

$24 = m \times 1.6$

$m = 15$ kg

(c) Neptune

g on Neptune is 12 N/kg and the acceleration due to gravity will be numerically the same as g. This is a problem solving question where a table of data has to be studied to reach the correct answer.

Credit question 2

The diagram shows a space shuttle consisting of an orbiter, called *Discovery*, and booster rockets. At lift off, *Discovery* and the booster rockets have a total mass of 2.05×10^6 kg and the thrust of the rocket engines is 2.91×10^7 N. The frictional forces acting on the shuttle at lift off are negligible.

At lift off:

(a) label the forces X and Y acting on the shuttle in the directions shown. (1)

(b) calculate the weight of the shuttle. (2)

(c) calculate the acceleration of the shuttle. (3)

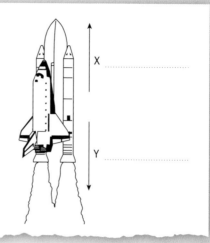

Credit question 2 – Answer

(a) X: thrust **or** engine force   Y: weight **or** force of gravity

(b) weight $W = mg = 2.05 \times 10^6 \times 10 = 2.05 \times 10^7$ N

(c) unbalanced force = thrust − weight = $2.91 \times 10^7 - 2.05 \times 10^7$

$$= 8.6 \times 10^6 \text{ N}$$

$F = ma$

$8.6 \times 10^6 = 2.05 \times 10^6 \times a$

$a = \dfrac{8.6 \times 10^6}{2.05 \times 10^6} = 4.2 \text{ m/s}^2$

A common mistake is writing the word **gravity** instead of **force of gravity** or **weight** in questions like this.

Remember that a calculation worth 3 marks will have more than one calculation.

Credit question 3

The International Space Station orbits Earth at a height of 380 km.

The command module of the space station has a mass of 20 tonnes (20×10^3 kg)

Masses as large as this are difficult to accelerate. Circle the term that is used for this concept:

gravitational field strength   **inertia**   **thrust**   **weight** (1)

Credit question 3 – Answer

inertia

Inertia as the command module is harder to start moving or accelerate, (and also harder to stop moving).
Questions on inertia are not very common – perhaps once every 5 years, so be prepared. Inertia refers to how hard or easy an object is to start moving or stop moving.

Credit question 4

During the Apollo 11 expedition to the Moon, 21 kg of soil samples were brought from the Moon to the Earth. The gravitational field strength was not constant throughout the journey.

(a) What is meant by gravitational field strength? (1)

(b) Complete the table to show the mass and weight of the soil samples at various stages of the journey. (3)

Credit question 4 – Answer

(a) *It is the force of gravity per kilogram.* **or** *Gravitational field strength is the force of gravity on every kilogram.*

> The majority of students omit the **per kilogram** part of the answer.

(b)

Stage	Gravitational field strength (N/kg)	Mass (kg)	Weight (N)
on the Moon	1.6	21	33.6
at a point during the journey	0	21	0
on the Earth	10	21	210

> The mass is constant at 21 kg at any location.
> The weight is calculated using $W = mg$.

Look out for

The mass of an object stays constant. Only the weight changes with location.

Projectiles

What you should know at General level...

A projectile is an object moving sideways as well as falling vertically. Projectiles move through the air with **curved paths**.

The projectile's curved path is caused by the **force of gravity accelerating it downwards**. At the same time the **horizontal speed stays constant** if air resistance is negligible.

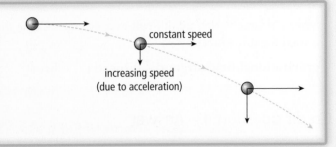

constant speed

increasing speed
(due to acceleration)

What you should know at **Credit** level...

Projectile motion can be treated as two independent motions. The horizontal motion can be treated separately from the vertical motion when solving numerical examples.

Horizontally the speed of a projectile stays **constant** as there are no horizontal forces acting on the projectile (air resistance is negligible).

Vertically the speed of a projectile increases as the projectile **accelerates** under the force of gravity.

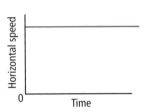

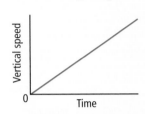

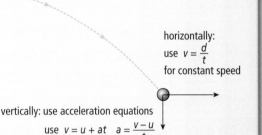

The equations used to solve projectile calculations are:

horizontally:
use $v = \dfrac{d}{t}$
for constant speed

vertically: use acceleration equations

use $v = u + at$ $a = \dfrac{v - u}{t}$

General question 1

In the passage below, circle one word or phrase in each set of brackets to make the statements correct:

A large stone and a small stone of the same material are kicked horizontally off a cliff at the same time. Both stones follow a **curved/straight/vertical** path.

Ignoring air resistance, the stones have the same **acceleration/mass/weight** because of the **force of gravity/force of friction/kick**. It is found that **the large/the small/neither** stone reaches the ground first. (4)

General question 1 – Answer

(curved)/straight/vertical

(acceleration)/mass/weight

(force of gravity)/force of friction/kick

large/the small/(neither)

This is a projectile motion with horizontal and vertical directions.

All masses will have the same acceleration.

The force of gravity causes the acceleration downwards.

Both stones have the same acceleration so reach the ground at the same time.

Credit question 1

An astronaut played golf on the Moon. The golf ball was struck horizontally from the edge of a steep crater. It landed 2 seconds later, 25 m away.

(a) Calculate the horizontal speed of the ball after being struck. (2)

(b) Calculate the vertical speed of the ball on landing. (2)

(c) How would the horizontal distance travelled by a ball projected with the same horizontal speed from the same height on Earth compare with that on the Moon. (3)

Credit question 1 – Answer

(a) $v = \dfrac{d}{t} = \dfrac{25}{2} = 12.5 \text{ m/s}$

(b) $v = u + at = 0 + 1.6 \times 2 = 3.2 \text{ m/s}$

(c) The horizontal distance on Earth would be smaller as the ball takes less time to hit the surface of the Earth and $d_{Earth} = v \times t$
= 25 × (a smaller value of time).

For vertical motion, use the acceleration equation $a = \dfrac{v - u}{t}$ rearranged to give $v = u + at$. The formula $v = u + at$ is not in the data booklet of formulae issued with the exam but is well worth remembering.
Also use $a = 1.6 \text{ m/s}^2$ as this motion takes place on the Moon.

This is a 3 mark question so think of 3 different parts to the answer. You must explain why the distance is less, not just state that the distance is less

Satellite motion

What you should know at **Credit** **level...**

Satellite motion is an extension of projectile motion. The faster a projectile is fired horizontally the greater distance it travels before reaching Earth's surface. The diagram shows a projectile being fired horizontally off a cliff at two different speeds. The faster speed will result in landing a greater distance from the base of the cliff.

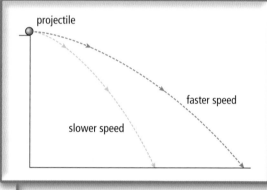

The surface of the Earth is flat on this diagram but it is actually curved because the Earth is a sphere. If the projectile has a fast enough initial horizontal speed then the shape of its curved path will be the same as the shape of the Earth's curved surface.

The projectile will now be falling to Earth but never reach the surface as the curved projectile path and the Earth's curvature are the same.

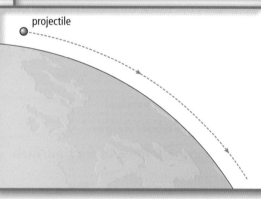

An artificial satellite is put in orbit by launching it from Earth then giving it the correct horizontal speed so that it falls back to Earth with the **same curved path as the Earth's curvature**.

The satellite motion must be above the Earth's atmosphere so there is no friction to slow down the satellite.

It is a common misconception that orbiting astronauts are weightless as there is no force of gravity acting on them. This is **not** correct. They appear to be weightless as they are falling back to Earth at the same rate as their satellite, not because the value of g is zero.

Credit question 1

A space probe goes into orbit around a planet

Explain why the probe follows a circular orbit. (2)

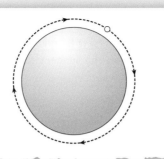

Credit question 1 – Answer

The probe accelerates towards the planet and has a steady speed forward. The projectile path has the same shape as the planet's curvature.

Satellite re-entry

What you should know at **General** and **Credit** level...

Artificial satellites which lose height will re-enter the Earth's atmosphere.

You have to know that **kinetic energy** will be **transformed** into **heat energy**, caused by **friction**, as the fast moving satellite comes into contact with the Earth's atmosphere.

You should be able to carry out numerical calculations involving the energy transformation (kinetic energy to heat energy) using the following relationships:

$$\tfrac{1}{2}\, mv^2 = cm\Delta T$$

or possibly

$$F \times d = cm\Delta T \quad \text{where a force of friction } F \text{ acts over a distance } d.$$

General question 1

When a spacecraft re-enters the Earth's atmosphere, some:

A heat is transferred to potential energy
B heat is transferred to kinetic energy
C kinetic energy is transferred to potential energy
D potential energy is transferred to heat
E kinetic energy is transferred to heat (1)

Heat is generated due to friction between the fast moving spacecraft and the Earth's atmosphere. This is an energy change question: always double check to see you don't have the energy names the wrong way round.

General question 1 – Answer

E

General question 2

In 1996 a spacecraft was launched on a mission to Mars. The spacecraft re-entered the Earth's atmosphere when it failed to break out of the Earth's orbit. Describe two effects on the spacecraft of the friction forces caused by air resistance on re-entry. (2)

General question 2 – Answer

Effect 1: spacecraft slows down   Effect 2: spacecraft heats up

Credit question 1

A spacecraft returning to Earth must re-enter the Earth's atmosphere. The spacecraft has a heat shield made from special silica tiles to prevent the inside from becoming too hot.

(a) Why does the spacecraft increase in temperature when it re-enters the atmosphere? (1)

(b) The mass of the heat shield is 3.5×10^3 kg and the gain in heat energy of the silica tiles is 4.7 GJ. Calculate the increase in temperature of the silica tiles. (3)

(c) Explain why the actual temperature of the silica tiles is less than the value calculated. (1)

Credit question 1 – Answer

(a) *Some kinetic energy of the spacecraft is turned into heat energy*

   caused by friction between the spacecraft and the air.

(b) C_{silica} = 1033 J/kg°C ◄

   $E_h = cm\Delta T$

   $4.7 \times 10^9 = 1033 \times 3.5 \times 10^3 \times \Delta T$

   $\Delta T = \dfrac{4.7 \times 10^9}{1033 \times 3.5 \times 10^3}$

   $\Delta T = 1300°C$ ◄

(c) *Heat energy is lost to the surroundings.*

This is a 3 mark calculation. You will have to look up the specific heat capacity of silica on the data sheet at the start of the exam paper.

C_{silica} = 1033 J/kg°C

GJ means gigajoule (10^9 J). A list of scientific prefixes and corresponding multiplication factors is given on the data sheet at the start of the Credit exam paper.

As was mentioned in the previous chapter this is a very common ending to a heat question. Remember that **to the surroundings** must be included.

It is not enough to put C for the temperature unit. The correct unit must also include the degree symbol ° to give °C.

Glossary

Acceleration (m/s²) the rate of change of speed i.e. change in speed divided by the time taken

Activity (Becquerels Bq) the number of atoms of a radioactive substance that disintegrate per unit time.

Aerial receives all incoming radio signals and converts them into electrical signals

Alpha particle a relatively large, slow moving radioactive particle with a positive charge (a helium nucleus)

Alternating Current (a.c.) the periodic movement of electric charges between the supply terminals

Ammeter an instrument for measuring electric current

Amperes (A) unit of electric current

Amplifier increases the amplitude of an electrical signal

Amplitude the peak value of an alternating quantity e.g. a wave

Amplitude modulation a carrier wave combines with an audio signal to produce an amplitude modulated wave for radio transmission

Analogue a signal that varies continuously

AND Gate a circuit with two inputs and one output in which the output signal is only high when both inputs are high

Angle of incidence the angle between a ray striking a surface and the normal to the surface at the point of incidence

Angle of refraction the angle formed by a refracted ray and the normal to the refracting surface at the point of refraction

Atom a chemical unit, composed of protons, neutrons, and electrons

Audio signal an electrical signal produced from the conversion of sound into electrical energy by a microphone

Average speed (m/s) the speed of an object over a relatively long period of time or distance

Battery two or more electric cells which produce electric charge

Background radiation nuclear radiation which affects us all of the time. It comes from the rocks, atmosphere and sun

Becquerels (Bq) the unit of acitivity of a radioactive substance where one becquerel is one atom decaying per second

Beta particle a relatively small, fast moving radioactive particle with a negative charge (electron)

Binary counter a circuit which counts binary pulses (a decoder is then used to convert these to decimal)

Brushes the parts of an electric motor used to connect the supply to the commutator and rotor coils

Capacitor an electronic component which stores charge

Carrier wave a high frequency wave which is combined with an audio signal for radio transmission

Chain reaction a neutron hits a uranium nucleus causing it to split (fission). This produces heat energy and more neutrons which in turn cause more fission to take place

Chemical energy energy produced from chemical compounds such as coal, oil and gas

Circuit breaker an electromagnetically operated device which can be used instead of a fuse to isolate household circuits

Clock pulse generator a circuit which uses a NOT gate and a resistor to produce a regular series of pulses

Commutator the contacts on the end of each rotor coil in a motor. They allow the current to change direction every half turn to allow continuous rotation of the motor

Conduction the method of heat travel in solids by the vibration of particles

Concave the shape of lens which causes light rays to diverge. Used to correct short sight

Convection the method of heat travel in liquids and gases where the particles are free to move

Convex the shape of lens which causes light rays to converge to a focus. Used to correct long sight

Critical angle the angle of incidence which causes an angle of refraction of 90°. This happens when light travels from a dense to a less dense material

Current (I) the flow of charge per unit time measured in amperes

Decibel (dB) unit used to measure the loudness of sounds

Decoder in radio receivers this separates the higher frequency carrier wave from the lower frequency audio signal. In the TV it separates the carrier wave from the video signal

Diffraction the bending of waves around a gap or obstacle

Diode an electrical component which allows current to flow through it in one direction only

Dioptre (D) a unit used to describe the power of a lens

Direct current (d.c.) the movement of charge which is always in one direction between the terminals of the supply

Discharge tube a type of lamp in which electrical energy is transferred to light in the gas

Double insulation where a device does not require an earth wire because the parts are protected by two layers of insulation

Earth wire a safety device to prevent electric shock. If a fault occurs it gives a low resistance path for the current causing the fuse to melt and isolate the appliance

Efficiency the ratio of useful work done by a system or device, to the energy supplied to it

Electrical energy type of energy associated with electric charge

Electromagnetic spectrum electromagnetic waves which range from long wavelength radio waves to gamma with short wavelengths. They all travel at the speed of light (3×10^8 m/s)

Energy (joules J) the ability of an object or system to do work

Equivalent dose (sieverts Sv) measures the biological risk of radiation

Eyepiece lens a short focal length convex lens which magnifies the image produced by the objective lens in a telescope

Field coils the coils around the outside of an electric motor used to provide a magnetic field which interacts with the field of the rotor coils

Filament lamp a type of lamp that uses a metal resistance wire to transfer electrical energy to light (and heat)

Fission the process of splitting the nucleus of an atom into smaller nuclei

Focal length the distance between the lens and the point where parallel light rays are brought to a focus

Force a pull or a push measured in Newtons. One newton is the force required to give a mass of one kilogram the acceleration of $1 \, m/s^2$

Fossil fuel a fuel which was formed from the remains of dead plants and animals over a very long period of time

Frequency (hertz Hz) the number of events per second or waves which pass a point in one second

Frequency modulation a carrier wave combines with an audio signal to produce a frequency modulated wave for radio transmission

Fuse an electric protection device in which a metal wire melts if too large a current passes through it

Gain the ratio of the output signal (voltage or power) to the input signal of an amplifier. It is a number and has no units

Galaxy a vast number of stars held together by a gravitational attraction

Gamma radiation electromagnetic radiation emitted from some radioactive nuclei

Geostationary satellite a satellite in an equatorial orbit which circles the Earth once every 24 hours so that it appears stationary above the surface of the Earth

Gravitiational field strength the force per unit mass exerted on an object by a planet. On earth it is $10 \, N/kg$

Gravitational potential energy energy due to height above the surface of a planet

Hertz unit of frequency. The number of events which happen per second e.g.waves per second

Half-life the time taken for the activity of a radioactive substance to decrease to half of its original value

Heat energy the form of energy which travels between two objects as a result of temperature difference

Image guide a light guide (opitical fibre bundle) in an endoscope which transmits light from inside a patient's body

image retention the eye and brain retain an image for a short time after a picture is removed from sight. This enables a series of still pictures to merge and appear to produce movement, e.g. on a TV screen

incident ray a ray of light which travels towards a surface from a light source

Infra-red radiation electromagnetic radiation with a longer wavelength than visible light. Commonly known as heat radiation

Input devices devices that produce a voltage or undergo a change in resistance in response to a change such as light level, temperature or force

Instantaneous speed the speed of an object at a particular time over a very short distance

Ionisation the process where an atom loses or gains an electron to become a positive or negative ion

joule (J) the unit of energy or work

Kilowatt-hours (kWh) the unit of energy used to measure electrical energy consumption in the home

Kinetic energy energy due to movement

Laser a device which produces a very concentrated, high energy, beam of light

Light Dependent Resistor (LDR) a resistor whose resistance decreases as the light level increases and vice versa

Light Emitting Diode (LED) an output device that only lights (conducts) when connected the correct way to the supply terminals. Usually requires a resistor in series to protect it from too large a current

Lighting circuit a type of household circuit used to connect lights in parallel

Light guide a bundle of optical fibres in an endoscope which transmits light into the patient's body to illuminate the internal organs

Light year the distance travelled by a beam of light in one year. Used by astronomers when measuring distances between stars

Line emission spectrum a unique set of wavelengths of light which each element emits. Used to identify elements in stars

Logic gates digital switching devices used in electronics e.g. AND gate, OR gate and NOT gate

Long sight a vision defect where close objects appear blurred and distant objects can be seen clearly

Loudspeaker converts electrical energy into sound energy

Mains supply general-purpose alternating current electric power supply. In Britain it has a declared value of $230 \, V$

Mass (kg) the amount of matter in an object measured in kilograms

Melting the change of state when a solid turns to a liquid

Microwaves electromagnetic radiation with wavelengths from 1 millimetre to 1 metre

Moon the Earth's only natural satellite

National Grid the high voltage electric cable network which carries electric current from power stations across the country

Newton (N) the unit of force usually abbreviated to N.

Newton balance a device used to measure the magnitude of a force

Newton pairs equal and opposite forces occurring when two objects interact. e.g. when object A exerts a force on object B there is an equal and opposite force exerted by object B on object A

Noise (decibels db) the name given to unwanted sound

Non renewable energy sources of energy which are not easily renewed and could one day run out

Normal a line drawn perpendicularly to the surface of an optical material and is used to define the angle of incidence of a light ray

NOT gate (inverter) a digital device which turns an input 1 into an output 0 and vice versa

Nuclear energy energy produced by the fission of radioactive nuclei

Objective lens a convex lens in a telescope with a long focal length and a large diameter. It collects light from a distant object

Ohm's law the formula linking current, voltage and resistance, $V = IR$

Open circuit a fault caused by a break or gap in a circuit

OR gate a circuit with two inputs and one output in which the output signal is only high when one input OR the other OR both are high

Oscilloscope an electrical instrument which displays a changing voltage as a two dimensional graph on a screen

Parallel circuit an electrical circuit where components are connected in parallel, so the current divides through each branch of the circuit

Planet a celestial body orbiting a star

Power (watts W) the rate at which energy is converted from one type to another

power gain the increase in power, for an amplifier, between input and output signals $P_{gain} = \frac{P_O}{P_i}$

Primary coil the input coil of a transformer

Principle of reversibility a light ray changing direction by reflection or refraction will follow the exact same path if the direction of the light ray is reversed

Radiation heat energy that travels in the form of waves. It does not need particles for transmission so can travel through a vacuum

Radio waves electromagnetic waves with the longest wavelengths and lowest frequency

Radioactivity the spontaneous disintegration or decay of atomic nuclei

Rating plate a label attached to electrical appliances giving information about voltage, and power etc

Real image made when light rays converge to form an image which can be seen on a screen

Reflected ray a ray of light after changing direction when coming into contact with a mirror or shiny surface

Refraction the change in speed (and often direction) when a wave enters a new material

Relay switch an electrically operated switch using an electromagnet to operate the switch

Renewable energy sources of energy which will not run out e.g. sunlight, wind, tides

Resistance (ohms Ω) a measure of an object's opposition to the flow of electric current

Ring main circuit the type of household circuit used for connecting the mains sockets

Rotor coil the coils in an electric motor which rotate

Secondary coil the output coil of a transformer

Semiconductor a material that conducts under certain conditions

Series circuit an electrical circuit where the components are connected one after the other so that there is only one route for the current

Short circuit a fault caused by a path of very low resistance created around a component e.g. two wires touching

Short sight a vision defect where distant objects appear blurred and close objects can be seen clearly

Sievert (Sv) the unit of equivalent dose

Solenoid a coil of wire used to produce a magnetic field

Solidifying the change of state from liquid to solid (the opposite of melting)

Specific heat capacity the amount of heat energy required to change the temperature of 1 kilogram of a substance by 1 degree Celsius

Specific latent heat of fusion the heat energy required to melt 1 kilogram of solid into liquid at the melting point of the substance. (Or the heat energy given out when 1 kg of liquid turns to a solid at its melting point.)

Specific latent heat of vaporisation the heat energy required to change 1 kilogram of a liquid into gas at the boiling point of the liquid. (Or the heat energy given out when 1 kg of a gas changes to a liquid at the boiling point of the liquid.)

Speed (m/s) a measure of how fast an object is moving

Star a massive ball of gases which produces heat and light

Streamlining changing the shape of a moving object to reduce the force of friction or air resistance on it

Sun the star at the centre of our Solar System

Temperature a measure of how hot or cold an object is, measured in degrees Celsius (°C)

Thermistor a resistor whose resistance decreases as temperature increases and vice versa

Thermometer an instrument to measure temperature

Total internal reflection (TIR) when light reflects off an inside surface of a transparent material with no refraction into any outer material

Tracer a radioactive substance injected into the body which can be monitored by external detectors as it passes through the body

Transistor a semiconductor device that can be used as a switch

Truth table a mathematical table which summarises the input and output variables of a logic gate

Tuner the part of a radio or TV which selects one particular frequency or wavelength

Turns ratio the number of turns on the secondary coil divided by the number of turns on the primary coil of a transformer or $\frac{n_s}{n_p}$

Ultrasound sound with a frequency greater than the upper limit of human hearing, i.e. 20 000 Hz

Ultraviolet radiation electromagnetic radiation with a wavelength shorter than visible violet light

Vaporisation the change of state from liquid to gas

Variable resistor a type of resistor whose value can be altered

Virtual image an image which cannot be projected onto a screen

Voltage (V) a measure of energy given to electric charges in a circuit

Voltage divider circuit a circuit made from two (or more) resistors in series, so that the supply voltage is divided across the resistors

Voltage gain the increase in voltage, for an amplifier, between input and output signals $V_{gain} = \frac{V_O}{V_i}$

Watts (W) the units of power. 1 watt is the same as 1 joule per second

Wavelength the distance between two consecutive crests of a wave and has the symbol λ

Weight (newtons N) the force on an object due to gravity

Weightlessness the sensation felt when there is zero force of gravity or when in free fall

Work done (joules J) the energy transformed when a force F acts over a distance d. *Work done = Fxd*

X rays electromagnetic radiation with frequencies between gamma rays and ultraviolet radiation

Formulae

$d = \bar{v}t$

$a = \dfrac{\Delta v}{t}$

$a = \dfrac{v - u}{t}$

$W = mg$

$F = ma$

$E_w = Fd$

$E_p = mgh$

$E_k = \dfrac{1}{2}mv^2$

$P = \dfrac{E}{t}$

$\text{percentage efficiency} = \dfrac{\text{useful } E_0}{E_i} \times 100$

$\text{percentage efficiency} = \dfrac{\text{useful } P_0}{P_i} \times 100$

$E_h = cm\Delta T$

$E_h = ml$

$Q = It$

$V = IR$

$R_T = R_1 + R_2$

$\dfrac{1}{R_T} = \dfrac{1}{R_1} + \dfrac{1}{R_2} + \ldots$

$V_2 = \left(\dfrac{R_2}{R_1 + R_2}\right)V_s$

$\dfrac{V_1}{V_2} = \dfrac{R_1}{R_2}$

$P = IV$

$P = I^2R$

$P = \dfrac{V^2}{R}$

$\dfrac{n_s}{n_p} = \dfrac{V_s}{V_p} = \dfrac{I_p}{I_s}$

$V_{gain} = \dfrac{V_0}{V_i}$

$P_{gain} = \dfrac{P_0}{P_i}$

$v = f\lambda$

$P = \dfrac{1}{f}$

$A = \dfrac{N}{t}$